CHEMICAL ENGINEERING METHODS AND TECHNOLOGY

HYDROQUINONE

PRODUCTION, USES AND HEALTH EFFECTS

CHEMICAL ENGINEERING METHODS AND TECHNOLOGY

Additional books in this series can be found on Nova's website under the Series tab.

Additional E-books in this series can be found on Nova's website under the E-book tab.

CHEMICAL ENGINEERING METHODS AND TECHNOLOGY

HYDROQUINONE

PRODUCTION, USES AND HEALTH EFFECTS

FRITZ GOKDEN
AND
ANDRÉ LAZZAROTTO
EDITORS

Nova Science Publishers, Inc.
New York

For permission to use material from this book please contact us:
Telephone 631-231-7269; Fax 631-231-8175
Web Site: http://www.novapublishers.com

Additional color graphics may be available in the e-book version of this book.

LIBRARY OF CONGRESS CATALOGING-IN-PUBLICATION DATA

Hydroquinone : production, uses, and health effects / Editors, Fritz Gokden and Andri Lazzarotto.
p. cm.
Includes bibliographical references and index.
ISBN 978-1-62100-258-1 (hardcover)
1. Hydroquinone. 2. Hydroquinone--Health aspects. 3. Chemical tests and reagents. I. Gokden, Fritz. II. Lazzarotto, Andri.
QD341.Q4H93 2011
661'.816--dc23
2011032381

Published by Nova Science Publishers, Inc. † New York

CONTENTS

Preface		**vii**
Chapter 1	Involvement of Cigarette Smoke-Related Hydroquinone in the Pathogenesis of Age-Related Macular Degeneration *Marianne Pons and Maria E. Marin-Castaño*	**1**
Chapter 2	Hydroquinone Solubility and Separation Processes *Fátima L. Mota, António J. Queimada and Eugénia A. Macedo*	**57**
Chapter 3	Permeation and Redox Mediation Processes at Poly(*O*-Aminophenol) Film Electrodes in the Presence of Hydroquinone/ *P*-Benzoquinone Redox Species: A Review Article *R. Tucceri, P. M. Arnal and A. N. Scian*	**103**
Chapter 4	Detection of Thiols Using Hydroquinone on Gold Surface *Luisa Rojas de Astudillo, Rosa Brito Gómez and Rolando J. Tremont*	**145**

Chapter 5	Cellular Effects of Hydroquinone *T. B. Ng, X. J. Ye* *and J. H. Wong*	**167**
Index		**185**

PREFACE

Hydroquinone has a variety of uses principally associated with its action as a reducing agent which is soluble in water. It is a major component in most photographic developers for film and paper, and is used as a topical application in skin whitening to reduce the color of skin. In this book, the authors present current research in the production, uses and health effects of hydroquinone. Topics discussed include the cellular effects of hydroquinone; the involvement of cigarette smoke-related hydroquinone in the pathogenesis of age-related macular degeneration; hydroquinone solubility and separation processes; the transport and transfer processes of poly(o-aminophenol)film electrodes in the presence of the hydroquinone/p-benzoquinone redox couple and the detection of thiols using hydroquinone on gold surfaces.

Chapter 1 - The goal of this chapter is to review the most relevant scientific evidence supporting the role of cigarette smoke-related hydroquinone in the pathogenesis of age-related macular degeneration (AMD), a major cause of blindness among the elderly and to highlight the importance of smoking cessation to reduce the risk for developing AMD. AMD is a late onset (after age 50), progressive degeneration of the retina associated with vision loss coupled with a spectrum of specific clinical, physiological and histopathological features. The early or dry stage of AMD is characterized by the accumulation of drusen, specific lipid rich deposits under the retina whereas the late or wet stage is due to the growth of abnormal new vessels under the retinal pigment epithelium RPE) from the subjacent choroid, termed choroidal neovascularization (CNV). Extracellular matrix turnover (ECM], inflammation, and angiogenesis are key cellular processes that play a central role in the pathogenesis of AMD. The scientific evidences to support the interpretation of the RPE injury by hydroquinone are reviewed in depth, especially the proposed role of hydroquinone on ECM turnover, inflammation, and CNV.

Chapter 2 - In chemical process design, separation unitsare usually very important for the purification of one or more components from the reaction stage. This step may involve different fluid or solid phases. Chemical engineering thermodynamics is the area of chemical engineering that studies the distribution of such components among equilibrium phases given temperature, pressure and feed composition. Therefore, reliable and accurate data on phase equilibria are crucial to optimize and scale up process separation units. As the number of real conditions in which the interest may be focused in is extremely high, thermodynamic models are frequently used to provide estimates of such data. Among these models, those based on the excess Gibbs energy, regularly known as activity coefficient models, and equations of stateare the most widely used. In this work, an overview of some thermodynamic models to describe the phase equilibria of hydroquinone and its isomers is provided, with the focus on the conditions commonly used for its separation processes. Results on the solid-liquid equilibria in water and organic solvents as a function of temperature, as well as in supercritical carbon dioxide as a function of pressure at several temperatures will be presented using the Non-Random Two Liquid Segment Activity Coefficient model (NRTL-SAC), the UNIFAC, the A-UNIFAC and the Cubic-plus-Association equation of state (CPA).

Chapter 3 - This review paper describes the different researches carried out in our laboratory to study transport and transfer processes at poly(*o*-aminophenol) (POAP) film electrodes in the presence of the hydroquinone/*p*-benzoquinone (HQ/Q) redox couple. The most significant experiments employing different techniques such as Cyclic Voltammetry, Rotating Disc Electrode Voltammetry and Electrochemical Impedance Spectroscopy are described in detail, and the application of existing models and theories that allow extracting charge-transport and charge-transfer parameters are also outlined. While an electron-transfer reaction (mediation reaction) occurs at negative potential values ($E < 0.0$ V *vs.* SCE), permeation processes are observed at positive potential values ($E > 0.8$ V *vs.* SCE). Freshly prepared poly(*o*-aminophenol) films were quantitatively deactivated to study how their permeation and charge-transfer processes are affected by the degree of deactivation (θ_c). While for low degrees of deactivation ($\theta_c < 0.4$) it is possible to recover the conductivity of a deactivated POAP film by reactivation in an alkaline solution, high degrees of deactivation lead to an irreversible deterioration of the polymer film. Electrochemical Impedance Spectroscopy applied at negative potential values allowed obtaining dependences of the different transport parameters on the degree of deactivation of the polymer film. While some parameters such as interfacial metal-film and film-solution resistances ($R_{m|f}$, $R_e^{f|s}$, $R_i^{f|s}$), the high-frequency capacitance (C_H) and the redox capacitance (C_p) exhibit a continuous variation without hysteresis between deactivation and reactivation processes within the whole θ_c range, others such as electron and ion diffusion coefficients (D_e, D_i) show not only marked changes of slope from given θ_c values but also hysteresis between consecutive deactivation and reactivation processes. The diffusion rates of the hydroquinone and benzoquinone species across the polymer film are also strongly reduced with the increase in the degree of degradation. This work could be interesting due to the wide range of potential applications of POAP. In this sense, although in practical applications of POAP it is necessary to maintain the conducting properties unaltered, the polymer is subjected to extreme conditions that can cause its partial deactivation. Thus, it seems to be important for electrochemists to know at least how the charge-transport process at POAP films changes with their deactivation.

Chapter 4 - Benzene, declared as an environmental contaminant, has commonly been used as an industrial solvent, a starting material for the synthesis of other chemicals, and as an antiknock agent in gasoline, and is a significant component of cigarette smoke. While not fully understood, the toxicity of benzene is linked to the production of reactive metabolites formed during cytochrome P4502E1. Benzene is metabolized by cytochrome P450 CYP2E1 to benzene oxide, which spontaneously forms phenol. Phenol is converted to hydroquinone by P4502E1. Hydroquinone and related metabolites are converted in the bone marrow, by myeloperoxidase, to benzoquinones, which are potent hematotoxic and genotoxic compounds that undergo further metabolic activation, by one-electron reducing enzymes, to generate semiquinones and reactive oxygen species (ROS). As an excess of ROS is produced, oxidative stress can occur and lead to altered cellular signaling. The production of ROS is tightly regulated and the cell possesses detoxifying mechanisms such as catalase, glutathione and superoxide dismutase to maintain appropriate cellular ROS levels. 1,4 benzoquinone (BQ) can alter signaling pathways and affect chromosomal integrity. Numerous attempts have been made to understand cytotoxic mechanisms influenced by the chemical structure of BQ generate DNA breaks and apoptosis. It is also a strong topoisomerase II poison. Its toxicity is mainly due to oxidative stress and/or forming Michael-type adducts with proteins and glutathione (GSH). Owing to the important physiological roles of quinones, such as hydroquinone,

1,4 benzoquinone, and reduced thiols such as GSH, it is necessary to develop sensitive and selective methods for the measurement of those substances to study cellular mechanisms of chemical-induced toxicity. For this purpose, the benzoquinone (BQ) is used in the electrochemical determination of thiol molecules. The Addition of sulfur nucleophiles to quinones has been studied and Characterized as 1, 4-reductive addition of the Michael type. Quinones, containing a polarized double bond, are considered Michael acceptors, where the Michael donors are thiol nucleophiles. The final product of this reaction is hydroquinone with a sulfur atom in the ortho position. If a metal surface is modified with a dithiol (to form a thiolate with the metal surface, leaving a free thiol,-HS) molecule, the reaction with BQ will be easy to produce hydroquinone (H2Q) moiety. The latter product can be used to determine electrochemically thiol-containing compounds in its structure.

Chapter 5 – Hydroquinone (HQ) is a major benzene metabolite that can be further oxidized toreactive quinones. It has been employed as a component of skin lighters and a photographic reducer.Its adverse effects on different cells have been demonstrated.High concentrations of HQ induces apoptosis in Jurkat cells. It induces DNA and chromosomal damage in mouse fibroblasts.It elicits DNA hypomethylation that accounts for the leukomogenicity of benzene.HQ causes apoptosis in human peripheral white blood cells. It induces oxidative damage anddiminishes expression of human 8-oxo-guanine DNA glycosylase mRNA in human A549lung adenocarcinoma cells. It hinders progression of T cell cycle by suppressingcyclin-dependent kinases. It brigs about alterations in protein expression inhuman hepatocytes. Exposure to HQ in utero causes a decline in the percentage ofdifferentiated HD3 avian erythroblast cells concomitant with a rise in reactiveoxygen species, both of which can be reduced by superoxide dismutase conjugatedto polyethylene glycol. Myeloperoxidase-activated HQ inhibits topoisomerase II,a,

In: Hydroquinone
Eds: F. Gokden and A. Lazzarotto
ISBN: 978-1-62100-258-1

Chapter 1

INVOLVEMENT OF CIGARETTE SMOKE-RELATED HYDROQUINONE IN THE PATHOGENESIS OF AGE-RELATED MACULAR DEGENERATION

Marianne Pons and Maria E. Marin-Castaño*
The Department of Ophthalmology, Bascom Palmer Eye institute,
University of Miami, Miami, Florida, US

ABSTRACT

The goal of this chapter is to review the most relevant scientific evidence supporting the role of cigarette smoke-related hydroquinone in the pathogenesis of age-related macular degeneration (AMD), a major cause of blindness among the elderly and to highlight the importance of smoking cessation to reduce the risk for developing AMD. AMD is a late onset (after age 50), progressive degeneration of the retina associated with vision loss coupled with a spectrum of specific clinical, physiological and histopathological features. The early or dry stage of AMD is characterized by the accumulation of drusen, specific lipid rich deposits under the retina whereas the late or wet stage is due to the growth of abnormal new vessels under the retinal pigment epithelium

* Reprint requests: Maria E. Marin-Castano, M.D., PhD. William L. McKnight Vision Research Center, 1638 N.W. 10th Avenue, Miami, FL 33136. Tel.: (305) 482-5142. E-mail: mcastano@med.miami.edu.

(RPE) from the subjacent choroid, termed choroidal neovascularization (CNV). Extracellular matrix turnover (ECM], inflammation, and angiogenesis are key cellular processes that play a central role in the pathogenesis of AMD. The scientific evidences to support the interpretation of the RPE injury by hydroquinone are reviewed in depth, especially the proposed role of hydroquinone on ECM turnover, inflammation, and CNV.

Keywords: Age-related Macular Degeneration, response to injury, reactive oxygen intermediates, blebbing, extracellular matrix turnover, actin cytoskeleton, Hsp25/27, MAPK, sub-RPE deposits, cigarette smoke, hydroquinone, inflammation, angiogenesis, animal model

Hydroquinone and Its Adverse Health Effects

Hydroquinone (benzene-1,4-diol) is an aromatic hydrocarbon metabolite of benzene (Figure 1) and ubiquitous environmental pollutant that has cytotoxic, haematotoxic, immunotoxic, and genotoxic properties [1-4]. It is a white, odorless, crystalline solid moderately soluble in water and highly soluble in alcohol. In the presence of water, hydroquinone can slowly oxidize to quinone, which is more volatile. Hydroquinone is used as a developer in the photographic industry, as an anti-oxidant in the rubber industry and as intermediate in the manufacturing of food antioxidants. It is also used as an ingredient in skin lighteners and is a natural ingredient in many plant-derived products, including fruits, grains, coffee, tea, beer, and wine. Following ingestion, hydroquinone is metabolized to water-soluble compounds in the liver, then filtered by the kidney, and eliminated in the urine. Exposure to hydroquinone from these common foods probably explains the presence of low concentrations of this quinone in the urine and plasma of people who have no occupational or other known exposure to hydroquinone.

Hydroquinone has also been identified in relatively high concentrations in the smoke of non-filtered cigarettes (up to 155 μg per cigarette) [5-7] and nominated for study by the National Cancer Institute based on its high levels of production and potential for human exposure [8]. In addition, hydroquinone is of interest as a metabolite of benzene via a cytochrome p450-mediated pathway [9,10] and as a model for quinone toxicity, since many chemicals and drugs of research interest contain the quinone nucleus [11].

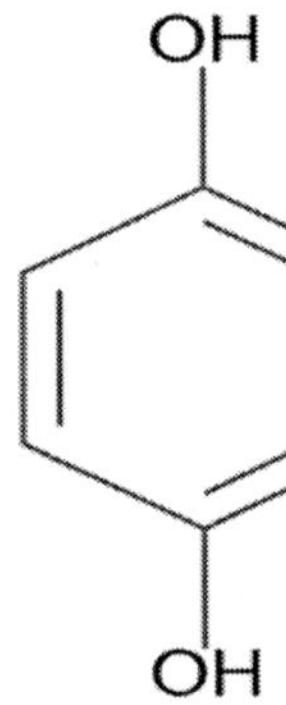

Figure 1. Chemical structure for hydroquinone.

All quinones can be futher bioactivated by myeloperoxidases and other hemo-protein perosidases [12] and become oxidants by a similar mechanism [7, 13, 14]. Hydroquinone competes with the normal substrate of mitochondrial oxidases associated with electron transport, and undergoes redox cycling with its corresponding semiquinone radical. As a result, hydroquinone metabolism in the mitochondria generates superoxide anion, which in turn, damages mitochondrial membranes and leaks into the cytoplasm [12]. In the cytoplasm, superoxide can be converted into hydroxyl anion and hydroxyl radicals. Intracellular hydroquinone-reactive oxygen species (ROS) generation participate in protein oxidation, lipid peroxidation [15], and DNA oxidation and damage [16], considering hydroquinone as the most potent DNA-damaging benzene metabolite and suggesting a role as strong pro-oxidant agent with chemical reactivity.

It has been reported that hydroquinone reacts with sulfhydryl groups of thiol compounds, such as L-cysteine, dithiothreitol, and N-acetyl-L-cysteine [17] and that its reactivity increases in the presence of nitric oxide (NO) [4]. This chemical reactivity appears to depend on the position of hydroxyl groups in the benzene backbone. Hydroxyl groups in the *para* position were confirmed to be the most reactive, since catechol with *ortho* groups and resorcinol with *meta* groups showed reduced inhibitory effects [17]. The effects of hydroquinone may thus occur as the result of direct binding to L-cysteine residues critically involved in the activity of target proteins.

Data from a number of laboratories provide compelling evidence that environmental toxicity of hydroquinone is an important consideration since previous reports have suggested the pathological importance of hydroquinone in the modulation of inflammation [17-19], immune responses [20-25], apoptosis , nephrotoxicity [26], skin allergy and pigment lightening [27-29], and eye irritation with impairment of vision among others [30, 31].

It has been shown that hydroquinone is toxic to immune-system and bone marrow cells [20, 21]. Thus, in chronic cigarette smokers, hydroquinone may cause increased rates of higher respiratory tract infection probably due to its immunosuppressive effects [20, 21]. According to Cho et al. hydroquinone suppresses most stages of innate immunity mediated by monocytes and macrophages, and of adaptive immunity by lymphocytes, suggesting that the initial activation of macrophages by bacterial infection might be diminished by hydroquinone exposure [22]. Additionally, hydroquinone can inhibit the migration of monocytes from the blood to infected areas and the aggregation of these cells at infection sites, and reduce the normal viability of lymphocytes from bone marrow and spleen [22]. Previous reports have also suggested that hydroquinone mediates allergic diseases via suppression of nuclear factor (NF)-κB binding activity and blockade of interferon IFN-γ and IL-12 production from activated T cells and macrophages respectively supporting an inhibitory role in Th1-type lymphocytes [23–25]. Hydroquinone is also able to cause renal tubular cell degeneration in the renal cortex [8] and induce apoptosis in human embryonic kidney cells through depletion of intracellular thiol [26], to destroy directly cytochrome P450 in human liver [32, 33], and to inhibit melanocyte tyrosinase in skin [34].However, it is important to note that because its potential cancer risk, some countries have banned the use of this compound as skin whitening. Hydroquinone has also been strongly implicated in producing leukemia associated with benzene exposure [35] and carcinogenesis [8]. Other human health effects related to hydroquinone are stimulation of nervous system [36], brown pigment deposition in the conjunctive and structural alterations of the cornea which can impair vision [30, 31]. As we will discuss below, cigarette smoke-related hydroquinone can damage the retina through oxidative processes thereby suggesting a role in the pathogenesis of age-related macular degeneration (AMD).

Age-Related Macular Degeneration

Age-related macular degeneration (AMD) is a late onset (after age 50), progressive degeneration of the retina associated with vision loss coupled with a spectrum of specific clinical, physiological and histopathological features [37, 38]. AMD affects 30% of people older than age 70, and is the leading cause of blindness in the elderly [39, 40]. Fourteen million people may be affected in the United States, and over 60 million worldwide. Since the population older than 60 years is the fastest growing segment in western society, the burden of AMD is becoming a huge global public health problem [39, 40].

The clinical presentation of AMD progresses through two stages: early and late [41]. Early AMD ("atrophic" or "dry" degeneration) is characterized by accumulation of drusen and other lipid rich extracellular deposits under the retinal pigment epithelium "(RPE) [41-43]. During aging, deposits initially accumulate between the RPE and its basement membrane (called basal laminar deposit or BLD), but progression into AMD requires additional deposit formation within Bruch's membrane (BrM), (called basal linear deposit (BLiD) and "nodular" drusen). Ultimately, early AMD can progress into late or severe atrophic AMD characterized by death of RPE and photoreceptors (called geographic atrophy) [41, 43] (Figure 2). Late AMD ("exudative" "neovascular" or "wet" degeneration) is characterized by endothelial invasion and pathological neovascularization under the retina [41, 43]. Wet AMD is always preceded by early disease [43]. Death or dysfunction of retinal photoreceptors is the ultimate cause of vision loss [44].

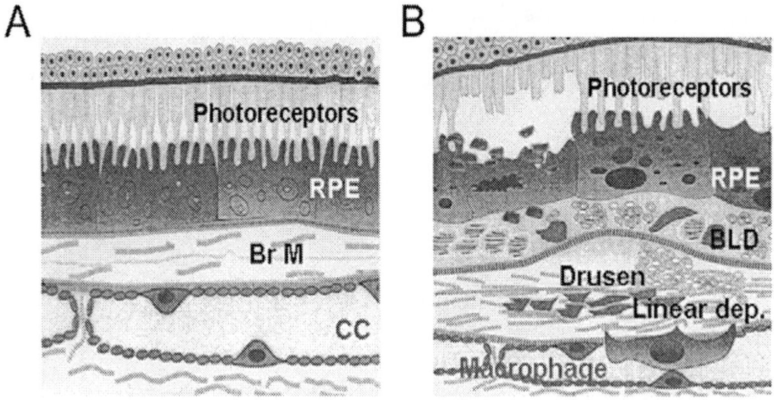

Figure 2. Normal anatomy of the retina (A). Histopathologic changes of early Age-related Macular Degeneration (AMD) (B). The schematic drawing shows the typical sub-Retinal Pigment Epithelium (RPE) deposits located in Bruch's Membrane (BM). Basal laminar deposits (BLD) are normally located between the RPE cell membrane and the RPE basement membrane while basal linear deposits (BLinD) and drusen are located in the inner collagenous layer of BM. Other histologic findings of this early stage of the disease are also shown such as devitalized photoreceptors and RPE cells, BM thickening, and invasion of macrophages. CC= Choriocapillaris.

However, the initial pathogenic target of AMD is the RPE, its extracellular matrix (called BrM), and the subjacent vascular bed [called choriocapillaris], the blood supply for the outer retina [39]. To date, little is known about the pathogenesis of dry AMD and events or factors that convert the dry form of the disease to its wet form. Treatment options currently available are very limited for this major public health concern. As a result, the development of preventive strategies is an area of great importance.

AMD is a multifactorial disease with age, systemic health, genetic and environmental risk factors influencing disease progression [45, 46]. The most important pathogenic factors leading up to AMD include oxidative stress, inflammation, and local production of angiogenic factors [47]. A substantial body of literature suggests a role for oxidant injury to the RPE and local inflammation as putative mechanisms in the pathogenesis of AMD [48-50]. Although intuitively obvious, oxidant injury can induce either lethal responses, leading to cell death, or nonlethal responses inducing a functional change from baseline compatible with continued life of the cell but leading to dysfunction of the tissue or organ. Most studies focus on oxidant-mediated death of RPE [51-54]. Yet, RPE death [so-called geographic atrophy] is a very late stage of dry AMD, resulting from a very chronic and progressive process which involves widespread atrophy of the RPE, inducing apoptosis of overlying photoreceptors and subsequent exposure of choroidal vessels. Subretinal deposits and thickening of BrM, the hallmarks of early AMD, develop decades before the RPE cells actually die. Therefore, nonlethal cellular responses to RPE oxidant injury must contribute to early AMD [6, 55].

Atrophic AMD is the most common form, affecting approximately 85% of persons with AMD. As outlined above the accumulation of specific deposits under the RPE (Figure 3) is a very prominent histopathologic feature of eyes with AMD [56-59]. The formation of drusen is an active process with distinct biological pathways. However, lack of scientific consensus exists regarding the origin of drusen. Research from different groups has been based on the conceptually innovative RPE injury hypothesis which proposes that deposit formation and accumulation is secondary to chronic, repetitive but nonlethal RPE injury RPE [60-66]. Irrespective of the injury, this model proposes that all stimuli result in a final common pathway of cellular responses that cause the actual deposits. Repetitive injury ultimately can kill RPE leading to late dry AMD [67].

More recently, inflammatory-derived injury stimuli have also been implicated, including oxidants, complement, immune complexes and factors produced by macrophages [47, 63, 68].

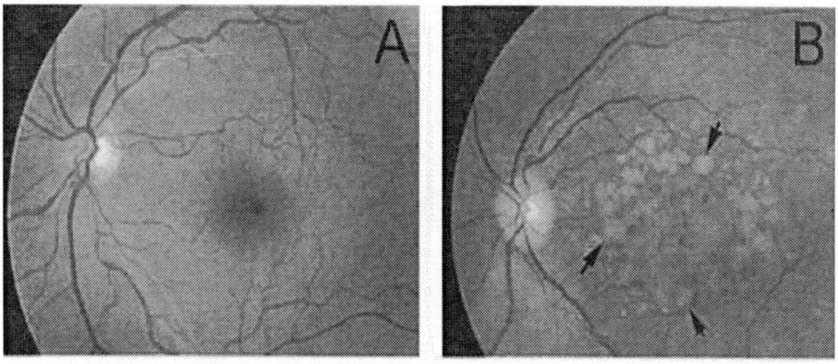

Figure 3. Color fundus photographs of a normal macula (A) and a macula with confluent soft drusen (black arrows) (B).

Inflammatory cells might be responsible for drusen progression into choroidal neovascularization (CNV) by secretion of growth factors and cytokines that will damage the choriocapillaris and stimulate the invasion of neovessels into the subretinal space [69-71]. Drusen may coalesce and progress, leading to atrophy of the surrounding retina. Moreover, its formation precedes and increases the risk of CNV, which characterizes the wet form of the disease. CNV is attributed to an unbalanced expression of angiogenic and anti-angiogenic factors, i.e. an alteration in the concentration of biomolecules that stimulate or inhibit growth of new blood vessels [72].

Pathogenic Mechanisms for Subretinal Deposit Formation

Although drusen were first described nearly 160 years ago, the cellular and molecular events involved in their formation have not been fully elucidated. Lack of scientific consensus exists regarding the origin of drusen, but at least five different paradigms are currently proposed to explain deposit formation in AMD; a) genetic hypothesis, b) lysosomal failure/lipofuscin hypothesis, c) choroidal hypoperfusion hypothesis, d) barrier hypothesis, and e) RPE injury hypothesis. Because our research is based on the RPE injury hypothesis, in this section this theory is reviewed in light of the more recent finding.

Some investigators have suggested that the incidence and severity of AMD has increased among populations exposed to western lifestyle and urban environments [73-75]. One possible explanation is that toxic substances associated with western lifestyle might directly contribute to the formation of drusen and late forms of AMD by an increase in oxidative stressors. Based on this idea, we reasoned that the redox molecule hydroquinone, a prevalent component of cigarette smoke, automobile exhaust and certain processed foods, might damage RPE contributing to drusen pathogenesis.

There are a number of direct and/or indirect lines of evidence supporting a role for the RPE in drusen biogenesis. According to traditional models of drusen formation, any cellular material residing within drusen is predicted to be of RPE origin. Indeed, RPE-derived basal laminae, organelles and cellular fragments, and even entire cells can be detected in early "drusen". Some authors have described the appearance of RPE "debris" blebbing into drusen or pre-drusen sites [76]. RPE constituents, such as basal laminae, as well as lipofuscin and melanin granules, are observed within early drusen, where they likely contribute to drusen volume and formation.

The theory that drusen were derived from damaged RPE was originally postulated by Donders, who first described drusen in a post-mortem eye, believed that drusen were derived from RPE nuclei, based on the supposition that the latter are relatively resistant to degradation [77]. Donders' theory was later modified by De Vicentis (1887) who proposed that degenerative change in the RPE cytoplasm, rather than in the nucleus, was the precipitating event. On the other hand, Muller (1856) proposed that drusen result from aberrant secretion of basement membrane components by the aged RPE [78].With the advent of electron microscopy, the substructural features of drusen were revealed, and new variants of the earlier theories were advanced. Some investigators have concluded, that drusen are formed when the RPE expels portions of its basal cytoplasm into BrM [76], possibly as a mechanism for removing damaged cytosol [79] or as a byproduct of phagocytic degradation [54]. Others have postulated that drusen are formed by autolysis of the RPE, due to aberrant lysosomal enzyme activity [80], although enzyme histochemical studies failed to demonstrate the presence of lysosomal enzymes in drusen [81]. Additional mechanisms for drusen formation, including lipoidal degeneration of the RPE [82], have been proposed.

In the modern version, the RPE injury hypothesis proposes that deposit formation is secondary to chronic, repetitive but nonlethal RPE injury [62, 63]. Two separate phenomena must be distinguished: the injury stimulus and the cellular response. The most widely implicated injury stimuli are various oxidants, especially those induced by RPE exposure to visible light or those derived from endogenous metabolism [64, 65]. Environmental oxidants derived from cigarette smoke, pollution or industrial by products have also been proposed to contribute to AMD [61, 62]. More recently, inflammatory-derived injury stimuli have also become implicated, including oxidants, complement, immune complexes and factors produced by macrophages or monocyte [68, 70, 71]. Inflammatory cells might be responsible for drusen progression into choroidal neovascularization by secretion of growth factors and cytokines that will damage the choriocapillaris and stimulate the invasion of neovessels into the subretinal space [69-71].

Irrespective of the injury, this model proposes that all stimuli result in a final common pathway of cellular responses that cause the actual deposits. Cellular responses that can lead to deposit formation include RPE cell membrane blebbing and dysregulation of extracellular matrix production and breakdown. Accumulation of sub-RPE blebbed material in the setting of imbalanced breakdown and resynthesis of basement membrane and BrM extracellular matrix is proposed to produce the various deposits of AMD. Repetitive injury ultimately can kill RPE, leading to late dry AMD [39].

Hypothetical Model for Dry AMD and Its Progression to Wet

As mentioned above, our research is based on the RPE injury hypothesis. We postulate that the pathogenesis and progression of dry macular degeneration is characterized by five distinct stages: [1] Initial RPE oxidant injury causes extrusion of cell membrane "blebs", together with decreased activity of matrix metalloproteinases (MMPs), promoting bleb accumulation under the RPE as BLD. Moreover, decrease in the production of pro-inflammatory monocyte chemoattractant protein-1 (MCP-1) by injured RPE leads to a decline in macrophage recruitment, reducing clearance of debris and BLD deposition; [2] when significant BLD will have already formed, RPE-derived blebs accumulated as BLD, which carried basigin and MMP-14, key MMP-2 regulatory molecules stimulate RPE basement membrane breakdown allowing the migration of BLD and buildup of BLiD or drusen; [3] RPE are subsequently stimulated to increase synthesis of MMPs, collagens and other molecules responsible for ECM turnover, affecting both RPE basement membrane and BrM. This process leads to the formation of new basement membrane under the RPE to trap these deposits within BrM. We postulate that various hormones and other plasma-derived molecules related to systemic health cofactors are implicated at this stage; [4] blebs and other debris accumulated as BLD activate MCP-1 secretion by RPE leading to recruitment of macrophages and/or dendritic cells to sub-RPE deposit areas, where they will scavenge, release inflammatory mediators, growth factors or other substances, which may promote complications and progression [69-71]; and [5] injured RPE cells increase vascular endothelial growth factor (VEGF) and decrease pigment epithelial derived factor (PEDF) proteins favoring angiogenesis and progression of drusen to CNV. As discussed below, the effect of hydroquinone in the mentioned stages will be reviewed in depth.

Association between Smoking and AMD

It has been postulated that environmental oxidants are frequently implicated in RPE injury and may contribute to deposit formation. Cigarette smoking is the strongest environmental risk factor for all forms of AMD, even in people exposed to passive smoking [83-86].

Plausible biological mechanisms such as direct oxidation by cigarette smoke components, depletion of antioxidant protection, immune system activation and atherosclerotic vascular changes, support the involvement of smoking in the etiology of AMD [87]. Although the pathogenesis of AMD and the mechanism of action of smoking on the eye are not fully understood [88], the risk of developing AMD is likely to involve more than one mechanism.

Cigarette smoke is comprised of a gas and tar phase. Each phase contains both inorganic and organic free radicals, including ROS, epoxides, peroxides, NO, nitrogen dioxide, peroxynitrite, peroxynitrates and various other free radicals [89]. Moreover, cigarette smoke contains a high concentration of potent oxidants such as acrolein [90-92], dioxin [93], Benzo(a)Pyrene [94], cadmium [95-97] quinones, nicotine, and NO [6, 98-100]. Many of them demonstrated to be toxic to ocular tissue affecting the eye through oxidative damage and abnormal vascularization, which are important factors in the mechanism(s) of AMD development.

Smoking habit appears to be related to the long-term incidence and progression of AMD [101]. Moreover, cigarette smoke depresses antioxidants (e.g. decreases plasma vitamin C and carotenoids), induces hypoxia and alters choroidal blood flow [102, 103]. Interestingly, the Age-related Eye Disease Study (AREDS) [104] found that reduction in plasma glutathione and cysteine oxidation correlated with benefit from anti-oxidant treatment for intermediate AMD [105]. Collectively, these studies, along with the studies on cigarette smoking, implicate oxidative stress as a mechanism of AMD. Genetic variations such as a susceptibility locus in or near the hypothetical LOC387715 gene were associated with AMD [106, 107]. It has been reported that this locus encodes a mitochondrial protein, raising suspicion for a role of the oxidative defense response in this disease. This locus is associated with smoking, and the combination of the LOC387715 polymorphism and smoking confers a higher risk for AMD than either factor alone [108, 109]. Moreover, chemical in cigarette smoke modifies oxidatively docosohexanoic acid (DHA), the most abundant fatty acid in photoreceptor tips, to carboxyethylpyrrole (CEP) [110] and other lipids which have been identified which "tag" oxidatively oxidized damaged photoreceptors in AMD [111]. In the RPE, multiple proteins isolated from lipofuscin are also oxidatively damaged by cigarette smoke compounds including malondialdehyde, 4-hydroxynonenal and advanced glycation end products (AGE) [112, 113]. Interestingly, AGEs accumulate in BrM including basal deposits and drusen, and CEP adducts appear in drusen isolated from AMD samples [114, 115].The ability to defend against oxidative stress by upregulating the anti-oxidant defense response is likely to be a pivotal event that mediates the initiation and progression of AMD. The molecular damage from oxidative modification illustrated here, suggests that the anti-oxidant response in the macula at some point, becomes unable to neutralize oxidative stress.

In addition to direct oxidative damage to tissue, oxidative free radicals from cigarette smoke can modulate the immune-inflammatory system in part, through enhanced expression of pro-inflammatory genes, as reviewed in Biswas and Rahman (2009) [116]. The discovery of polymorphisms in several complement factors with AMD susceptibility points toward a specific role for complement mediated inflammation in the pathophysiology of AMD [117, 118]. AMD has been associated with local inflammatory responses in the RPE/choroid [119]. In previous works, it was demonstrated that the aged RPE/choroid becomes immunologically active [120] due to increased expression of complement components. Smoking also influences the function of the alternative pathway of complement activation [121]. Smoking seems to alter the C3 component of the complement and to reduce the efficacy by which it binds to complement factor H (CFH).

The association between smoke and ApoE has also been investigated. Smokers are known to have higher serum cholesterol and LDL levels, a possible risk factor for AMD development [122]. Impaired cholesterol metabolism in particular ApoE isoforms carriers could be further enhanced among smokers. Moreover, smoking is associated with decreased endothelial NO production, while ApoE-4 genotype has been reported to increase NO synthesis, relative to ApoE2 and ApoE3 [123].Thus, the presence of the ApoE2 protein in smokers may reduce endothelial NO levels in cell tissues. Given that the normal function of NO in the retina is to neutralize circulating oxidized lipids, its decreased availability may promote oxidative damage to the RPE cells.

As mentioned above, quinones, acrolein, BenzoPyrenes, and cadmium can damage the eye through oxidative processes. It has been demonstrated that hydroquinone the most abundant quinone in cigarette smoke causes oxidative damage, apoptosis, increases lipid peroxidation and mitochondrial superoxide production, and decreases intracellular glutathione in the RPE cells [124]. Acrolein, a product of lipid peroxidation in vivo, is a mitochondrial toxicant in RPE cells, which induces oxidative mitochondrial dysfunction and oxidative damage to proteins and DNA causing loss of cell viability [90-92]. Whereas Benzo(a)Pyrene causes mtDNA damage, alteration in the lysosomal activity, complement activation in the aged RPE, and upregulates expression of complement pathway components such as C3a, C5, C5b-9, and CFH in the RPE/choroid contributing to drusen formation [126, 127]. Cadmium [Cd], is another toxic which is concentrated in the body through cigarette smoke. Cd has been demonstrated to be a potent inflammatory, an oxidant agent, and accumulates with aging, effects which are linked to AMD [97, 97, 127]. In vitro studies have revealed Cd as an important potential factor in RPE cell death associated retinal disease [96].

Cigarette smoke also induces angiogenesis promoting CNV and progression to neovascular AMD. Nicotine is one of the major components of the cigarette and has been demonstrated to enhance the generation of radical oxygen species producing oxidative stress [128-130] and promote the generation of pro-inflammatory responses leading to chronic inflammation in smokers [131]. Futhermore, nicotine is responsible for the mitogenesis of endothelial and smooth muscle cells [132] regulating abnormal responses to vascular injury [133] and endothelial functions [134]. We and other previously showed that nicotine by its direct action on nicotinic receptors promoted angiogenesis and increased size and severity of CNV in a mouse model [133, 135]. Moreover, it has been shown that nicotine targets the retinal microvasculature by reducing the apoptotic rate of vascular endothelial cells and inducing the formation of new capillarity, mediated at least in part by release of vascular endothelial growth factor (VEGF) [134], which is one of the key growth factors involved in neovascularization and implicated in the most severe form of AMD [136-139]. Nicotine has also the ability to stimulate the growth of smooth muscle cells after oxidative injury, leading to the development of CNV in the retina. Moreover other cigarette smoke components such as dioxin induce expression and release of VEGF by RPE cells promoting CNV [124, 140].

Cell Membrane Blebbing and Extracellular Matrix Turnover: Role of Hydroquinone

A substantial body of literature suggests a role for oxidant injury to the RPE as a putative mechanism in the pathogenesis of AMD and addresses the protective actions of anti-oxidant. Although intuitively obvious, oxidant injury can induce either lethal responses, leading to cell death, or nonlethal responses inducing a functional change from baseline compatible with continued life of the cell but leading to dysfunction of the tissue or organ. Most studies focus on oxidant-mediated death of RPE [141-144]. Yet, RPE death (so-called geographic atrophy) is a very late stage of dry AMD, resulting from a very chronic and progressive process. Subretinal deposits and thickening of BrM, the hallmarks of early AMD, develop decades before the RPE cells actually die. Therefore, nonlethal cellular responses to RPE oxidant injury must contribute to early AMD.

We embrace the pathogenic paradigm based on the response-to-injury hypothesis which proposes that sub-RPE deposits originate from RPE-derived cell membrane blebs and dysregulation of ECM turnover induced by chronic nonlethal injury to the RPE in response to oxidative damage and propose that cigarette smoking-related hydroquinone contributes to the pathogeneis of AMD by causing oxidative damage to the RPE.

Nonlethal Cell Membrane Blebbing

Nonlethal cell membrane blebbing is the process by which a cell can pinch off part of its plasma membrane and cytosol in an attempt to discard damaged cellular organelles, molecules, and lipid membrane [145, 146] and was first introduced 30 years ago as a possible pathogenic mechanism in drusen formation [76, 79, 147, 148]. Blebbing is an early morphologic sign of cell injury which occurs immediately after exposure to a wide variety of toxic agents. However this process is different from lethal blebbing and apoptosis [149-151].

Blebbing are closely interrelated with the actin cytoskeleton [152, 153-160]. The cytoskeleton is a dynamic structure undergoing continuous turnover by disassembly and reassembly of G-actin monomers, which are added to the active edge of a growing filament. Cytoskeletal turnover is mediated by actin polymerizing protein system (APP), which is a large system of interacting proteins.The identity of these proteins is a rapidly growing field, and includes kinases, phosphatases, cleavage proteins, and elongating proteins. Their individual specific function and regulatory interactions remain essentially unknown. However, in general, the APP functions as a coordinated system such that individual molecular regulators of cytoskeleton dynamics exert their actions on the entire APP in coordinated fashion [152, 153-160].

Very little information is available regarding the molecules that regulate the APP system and disease pathogenesis. Interestingly, oxidant injury is an effective activator of increased cytoskeleton turnover, presumably secondary to direct oxidative changes of disulfide bonds within actin filaments [152, 161-164]. The best characterized inhibitor/ stabilizer of cytoskeleton turnover is heat shock protein 25/27 (Hsp25/27). This molecule belongs to the family of small heat shock proteins [which includes α-β crystallins] and serves a dual role as cytoskeleton stabilizer and chaperone protein [163-168].

Oxidant injury can induce a rapid upregulation of Hsp25/27 protein through the activation of presynthesized heat shock factors (hsf-1,-2,-3) which upon oxidation form trimers that become very potent transcription factors [156-161]. Upon phosphorylation, Hsp25/27 acts as an F-actin cap-binding protein, binding to the active edge of a growing microfilament, causing a transient delay in actin cytoskeleton reassembly and repair [152, 153-170]. Transient delay in reassembly allows the unsupported plasma membrane to be forced outward by the hydrostatic forces of the cytoplasm, and a vesicle or bleb forms if cortical cytoskeleton and microfilaments do not completely reform within the outpouched membrane.

Phosphorylation of Hsp25/27 is mediated through the MAP kinase cascade, but is directly phosphorylated by MAPKAP 2 kinase (MK2), which in turn is phosphorylated by p38 kinase (Figure 4). [152;171] Thus, oxidative stress induces profound rearrangements of the actin cytoskeleton [151, 161, 172] leading to membrane blebbing through the activation of p38 mitogen activated protein kinase (MAPK)//Heat shock protein 27 (Hsp27) pathway [162, 172]. In response to stress, phosphorylated Hsp27 undergoes conformational changes and reorganizes into dimeric units [172, 173-176]. Phosphorylated Hsp27 regulates actin filaments dynamics by repressing the ability of Hsp27 to block actin polymerization [175].

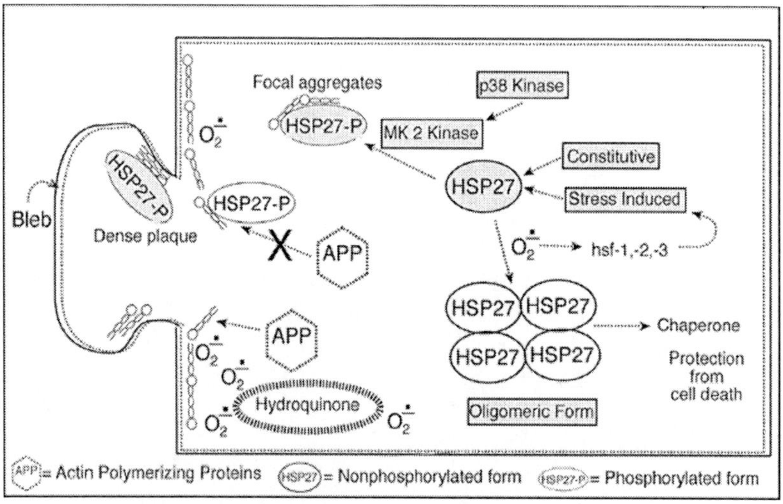

Figure 4. Regulation of the actin cytoskeleton turnover and its relationship to blebbing. The schematic drawing shows the phosphorylation of Hsp25/27 through the MK-2 kinase pathway and the capping of actin filaments which prevents the reassembly of the growing actin microfilaments that make the cytoskeleton in the cytoplasm by interfering with the function of the actin polymerizing proteins (APP) system and to disassembled aggregates in damaged cell membrane (i.e., dense plaques) after oxidant injury. Oxidant injury also causes Hsp25/27 to polymerize into large oligomers after which it becomes a potent chaperone molecule capable of binding to other proteins to protect them from injury stimuli, by promoting protein folding, and scavenging cytoplasmic oxidants. Also, oxidant injury can induce a rapid upregulation of Hsp25/27 protein through the activation of presynthesized heat shock factors (hsf-1,-2,-3) which upon oxidation form trimers that become very potent transcription factors.

Hsp27 phosphorylation has been abundantly described in several human diseases [177]. Yet, there is a complete lack of information regarding the possible association between phosphorylated Hsp27 and AMD. Increased Hsp27 protein content along with evidence of cellular oxidative stress was reported in human eyes with AMD, but Hsp27 phosphorylation was not investigated [178].

Therefore, we believe that Hsp27 is an important mediator of RPE response to hydroquinone-induced oxidative damage, which may contribute to injury-induced actin rearrangement and blebbing. We studied the phosphorylation of Hsp27 in RPE from human eye donors and found that the RPE constitutively express phosphorylated Hsp27, and that its expression is increased in patients with dry AMD (Figure 5) [179] providing novel evidence that phosphorylated Hsp27 may play a major role in the pathogenesis of AMD.

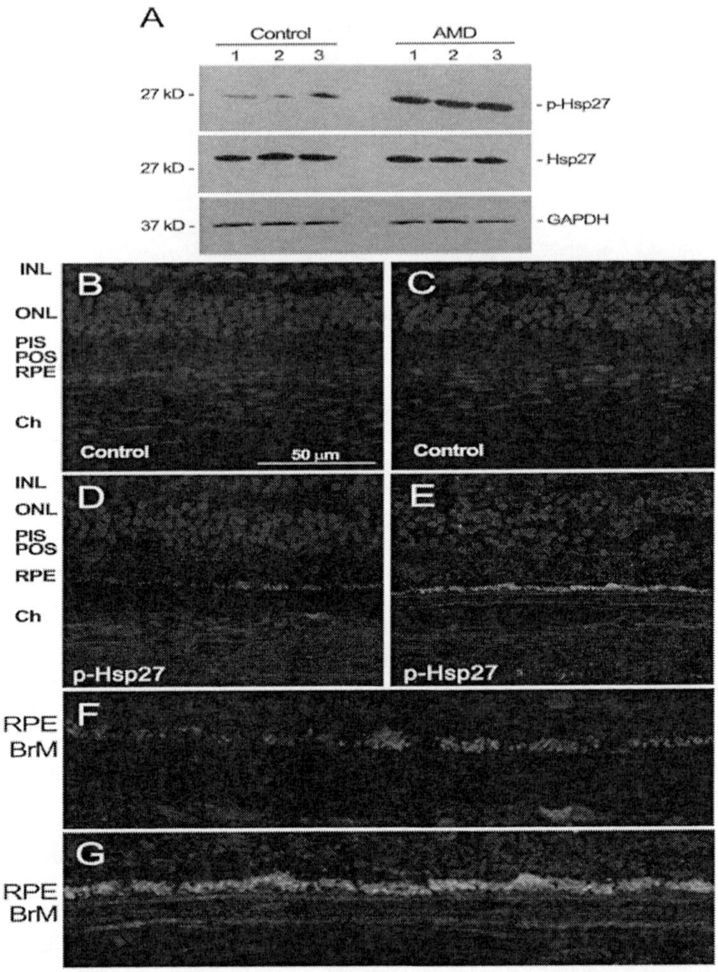

Figure 5. Increased phosphorylated Hsp27 (p-Hsp27) expression in human RPE from patient donors with dry AMD. A: Representative Western blot for p-Hsp27, total Hsp27 and GAPDH on RPE lysates from 3 donors with dry AMD and 3 controls with no known history of eye disease. B-G: Representative immunofluorescent double staining of p-Hsp27 (green) and nuclei (bleu) in retina sections from human donor eyes with no known eye disease (D), and human donor eyes with dry AMD (E). Negative controls were generated by omission of the primary antibody (B, C). Higher magnification showing RPE and BrM in control (F) and AMD (G) patients. Sections were analyzed using confocal microscopy (magnification x40 and x400). INL, inner nuclear layer; ONL, outer nuclear layer; PIS, photoreceptor inner segments; POS, photoreceptor outer segments; RPE, retinal pigment epithelium; Ch, choroid; BrM, Bruch's membrane.

Using a genetically modified human RPE cell line containing fluorescent protein anchored to the inner leaflet of the plasma membrane, we reported that human RPE cells constitutively express high levels of Hsp27 which were regulated by oxidative-mediated injury and that Hsp27 is expressed in extruded blebs confirming that Hsp27 plays a role in actin filaments dynamics [179, 180]. We also addressed the question whether hydroquinone-induced oxidative injury can activate different MAPK pathways of any posttranslational modifications such as dimerization or phosphorylation of Hsp27 known to regulate F-actin polymerization. Our previous observations were extended by showing that exposure of RPE cells to hydroquinone non lethal injury induces the transcriptional activation of Hsp27, accumulation of Hsp27 dimers and a rapid phosphorylation of Hsp27 as well as actin rearrangements inton aggregates and membrane blebbing (Figure 6) [179].

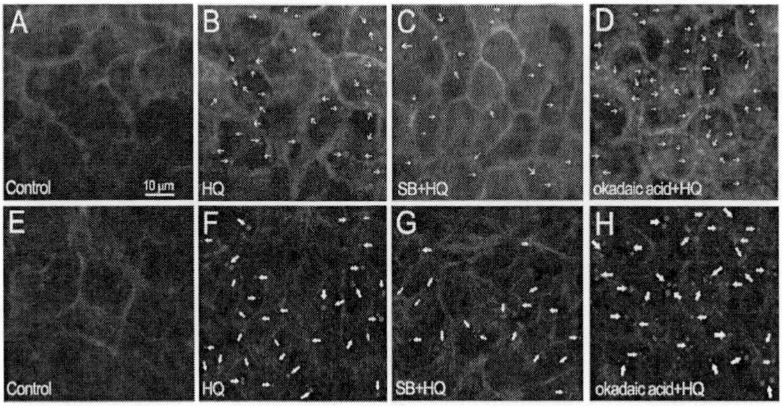

Figure 6. Induction of cellular changes in GFP-ARPE-19 cells exposed to hydroquinone (HQ). Membrane blebbing (A to D) and formation of F-actin aggregates (E to H). Confluent GFP-ARPE-19 cells were pretreated with or without 20mM of SB203580 (SB) or 60nM of okadaic acid, and then exposed to 100µM HQ for 6 hours. Cells were directly observed under a fluorescence microscope (top panel) or stained with rhodamin-phalloidin and then examined under a fluorescent microscope (bottom panel). A: Control GFP-ARPE-19 cells in which GFP was localized to the membrane. Membrane-derived blebs (thin arrows) after exposure to 100µM HQ for 6 hours (B),20µM of SB for 1 hour followed by 100µM HQ for 6 hours (C), and 60nM of okadaic acid for 1 hour followed by 100µM HQ for 6 hours (D). E: Control GFP-ARPE-19 cells showing normal F-actin filaments. F-actin aggregates (thick arrows) after exposure to 100µM HQ for 6hours (F), 20µM of SB for 1 hour followed by 100µM HQ for 6 hours (G), and 60nM of okadaic acid for 1 hour followed by 100µM HQ for 6 hours (H). Images are representative of three independent experiments. Magnification is x400.

These results together with the observation that SB203580, a specific pharmacologic inhibitor of p38 kinase activity [181], efficiently blocked Hsp27 phosphorylation as well as actin cytoskeleton remodeling and blebs formation in response to hydroquinone in RPE cells, strongly suggest that p38 MAPK pathways activation by hydroquinone modulates F-actin aggregates formation and membrane blebbing through Hsp27 phosphorylation.

Our findings are in agreement with prior studies reporting p38 MAPK signaling pathway as an upstream mediator in oxidative stress-induced actin reorganization and Hsp27 phosphorylation [162, 172-174, 182].

Several reports have shown that PP2A is involved in Hsp27 dephosphorylation [183, 184] and that oxidative stress causes extracellular-signal-regulated kinases [ERK] phosphorylation and reorganization of actin cytoskeleton in RPE cells [185].

Using okadaic acid as an inhibitor of Hsp27 dephosphorylation by PP1 and 2A, we observed an increase in Hsp27 phosphorylation in hydroquinone-stimulated RPE cells as well as F-actin reorganization and blebs formation [179] [Figure 6]. Moreover, we demonstrate that Hsp27 phosphorylation and F-actin aggregates formation is almost completely abolished in cells transfected with siRNA against Hsp27 following treatment with hydroquinone [179].

As mentioned above ERK cascade participates in numerous intracellular signaling pathways in response to environmental stimuli, such as oxidative stress [185].

Our study also show that treatment with hydroquinone led to a robust activation of ERK signaling pathway in ARPE-19 cells as well as in mice. These results together with the observation that PD98059, a specific pharmacologic inhibitor of MEK, completely abolished Hsp27 phosphorylation as well as actin cytoskeleton remodeling in response to hydroquinone, strongly suggest that ERK is also a key upstream activator of hydroquinone-induced Hsp27 phosphorylation in RPE cells [179] (Figure 7).

Our results not only showed that kinetics of p38 and ERK phosphorylation correlated well with that of Hsp27, but also that Hsp27 phosphorylation and F-actin aggregates formation were decreased after inhibition of either p38 or ERK signaling cascades.

These observations suggest that p38 as well as ERK MAPK pathways are required for the optimal activation of Hsp27 leading to F-actin rearrangement and bleb formation in RPE cells in response to hydroquinone (Figure 6).

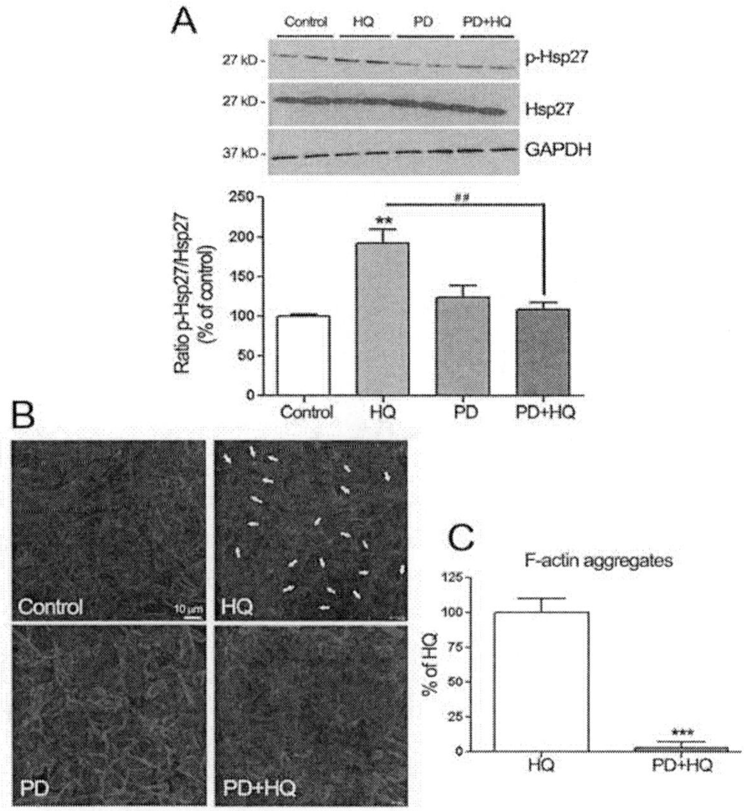

Figure 7. Inhibition of ERK MAPK pathway blocks hydroquinone (HQ)-induced Hsp27 phosphorylation and focal aggregates formation. A: Abrogation of HQ-induced Hsp27 phosphorylation by PD98059. Confluent serum-starved ARPE-19 cells were pretreated for 1 hour with 40μM of ERK inhibitor PD98059 (PD), and then exposed to 100μM HQ for 5 min. Total Hsp27, p-Hsp27 and GAPDH protein expression was evaluated by Western blot analysis. Top: Western blot from a representative experiment. Numbers on the left represent protein molecular mass in kilodaltons (kD). p-Hsp27 protein expression was normalized to total Hsp27 protein. Bottom: average densitometry results of three independent experiments run in duplicate. Data are expressed as percentage of control and are means ± SE. $^{**}p < 0.01$ versus control; $^{\#\#}p < 0.01$ versus HQ alone. B: Decreased formation of F-actin aggregates showed by staining for F-actin in ARPE-19 cells pretreated for 1 hour with or without 40μM of PD and then exposed to 100μM HQ for 6 hours. Cells were stained with rhodamine-phalloidin and examined by confocal microscopy using magnification x40. White arrowheads show formation of focal aggregates. C: Quantification of F-acting aggregates from three independent experiments run in duplicate. Data are expressed as percentage of HQ-treated cells and are means + SE. $^{***}p < 0.001$ versus HQ-treated cells.

Taken together,these data establish; a] a direct correlation between levels of phosphorylated Hsp27 and actin cytoskeleton reorganization in response to hydroquinone-induced oxidative injury in human RPE cells; b) present ERK as a novel upstream positive regulator of Hsp27 and actin aggregates formation in response to hydroquinone-induced oxidative injury in RPE cells; and c) give support to a key role of phosphorylated Hsp27 in the regulation of F-actin filaments dynamics and blebs formation following hydroquinone-induced oxidative stress in RPE cells. Given that there is no effective treatment for dry AMD, this study highlights Hsp27 as a potential, disease-related protein as well as biochemical pathways for potential therapeutic strategies.

Extracellular Matrix Turnover

Another injury response relevant to early AMD is imbalance in extracellular matrix (ECM) turnover. The normal anatomy and physiology of ECM in most tissues requires continuous turnover of matrix components by a tightly regulated balance in the production of matrix molecules like collagen, laminin, matrix metalloproteinases (MMPs), and tissue inhibitors of metalloproteinases (TIMPs) [186, 187]. It has been shown that relatively small dysregulation in the relative production of MMPs, TIMPs, and collagen types I and IV [186, 187] may lead to net changes in the ECM, including thickening and deposit formation [187, 188]. Accordingly, dysregulated turnover of ECM is a major mechanism of disease pathogenesis in many tissue sites, including renal disease, atherosclerosis, lung disease, and others [187-190]. Unfortunately, minimal information is available concerning normal turnover in healthy BrM or imbalanced turnover in AMD.

Ample evidence supports the idea that MMPs and their tissue inhibitors TIMPs play an important role in AMD. The RPE is capable of producing many ECM components such as MMP-2, MMP-14, Basigin, also known as EMMPRIN (extracellular matrix metalloproteinase inducer) or CD147, collagen, and TIMP-2 [60, 61, 191, 192]. MMP-2 and MMP-14 are the major RPE enzymes synthesized for the degradation of matrix type IV and I collagens, laminin, and fibronectin, which are components found in BrM [191, 193]. MMP-2 is the key enzyme for ECM turnover in BrM and is synthesized as an inactive zymogen pro form (pro-MMP-2) [186]. The transmembrane metalloproteinase MMP-14 is well known to activate MMP-2 in a specific manner [193]. On the other hand, TIMP-2 may inhibit the cleavage of pro-MMP-2 into MMP-2 [187]. Based on these data our research focuses on

MMP-2, MMP-14, basigin, and TIMP-2 and propose that dysregulation of MMP-2 is the primary cause for the accumulation of sub-RPE deposits that become BLD and drusen. Thus, preservation of RPE-derived MMP-2 function will prevent these events.

As mentioned before, oxidative injury to the RPE may not only produce blebbing but also dysregulated ECM turnover [40, 62, 64, 180, 194-196]. Our group has demonstrated that nonlethal oxidant injury to the RPE with hydroquinone induces a wide range of changes in gene expression, especially for those genes involved in regulation of ECM [63], and that blebs form and accumulate in the absence of activated MMP-2, which otherwise induce breakdown of types I and IV collagen present in the basement membrane of the RPE [196]. Sustained oxidant injury with hydroquinone is capable of downregulating MMP-2 activity in an in vitro system of RPE cells, correlating with an increase in bleb levels [60, 61]. Also, type IV collagen accumulation, the main component of the RPE basement membrane, correlated with the absence of MMP-2 activity [61]. Accordingly, we have shown in vivo [196] that reducing MMP-2 activity leads to an increase in deposit formation by RPE cells. This allows the blebs to become trapped between the RPE cell membrane and the basement membrane to form BLD.

For the BLD deposits to progress into linear deposits ECM upregulation with MMP-2 activation is necessary for collagen and BLD deposit degradation. This will ultimately displace sub-RPE deposits into the inner layer of BrM where BLinD and drusen are histologically found. As proposed in our hypothetical model fro dry AMD, RPE-derived blebs accumulated as BLD could stimulate RPE basement membrane breakdown allowing the migration of BLD and buildup of BLiD or drusen. We postulate that various plasma-derived molecules related to systemic health cofactors (i.e., Angiotensin II) are implicated in this stage.

As an attempt to better understand the mechanisms involved in early AMD, we investigated the proteomic profile of RPE-derived blebs induced by hydroquinone. Interestingly MMP-14 and basigin were identified in the RPE blebs [192].

Both proteins are of special relevance to the AMD pathology as they promote MMP-2 activation. Basigin, is a pleiotropic transmembrane glycoprotein [197, 198]. The basigin gene (BSG) encodes for a 29 kDa protein which is prone to glycosylation increasing its molecular weight between 35 and 65 kDa [199-201]. Induction of matrix MMPs constitutes the most relevant function of basigin [201, 203-205]. A clear requirement for basigin to be glycosylated exists for MMP induction [206-208].

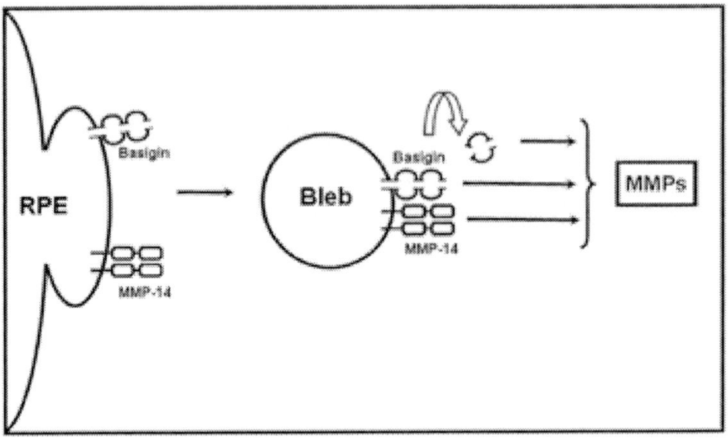

Figure 8. Mechanistic model by which basigin and MMP-14 are carried to distal sites from RPE where they exert their actions promoting MMP-2 activity.

The fact that basigin is a prominent MMP inducer provides this protein with a putative role in normal tissue remodeling physiology and ECM pathologies other than cancer. In particular, glycosylated basigin has been reported to be necessary for maturation of the retina photoreceptor cells [209]. Within the mature eye, basigin is expressed in the RPE, Mueller cells, and in endothelial cells of blood vessels [210-213].

A known mechanism by which basigin increases the MMPs occurs at transcriptional level [214, 215]. However, shedding of basigin from the plasma membrane constitutes an additional regulatory mechanism recently proposed where basigin is transported in microvesicles which are later on degraded, releasing soluble, active basigin [216].

This proposed mechanism may permit basigin to exert its actions at distant sites. MMP-14 is an interesting candidate responsible for the basigin shedding [217-219]. Based on the cited evidence, we hypothesize a mechanistic model by which basigin and MMP-14 are carried to distal sites from the RPE where they exert their actions promoting MMP-2 activity. In addition, MMP-14 may interact with basigin releasing fully functional "soluble" basigin (Figure 8).

We reported the presence of glycosylated basigin and MMP-14 in blebs, which was confirmed by Western blot and immunofluorescence staining in human control and AMD retinas (Figure 9) [192]. Basigin and MMP-14 were confined to the RPE in normal retina, while it was widely expressed in AMD retina.

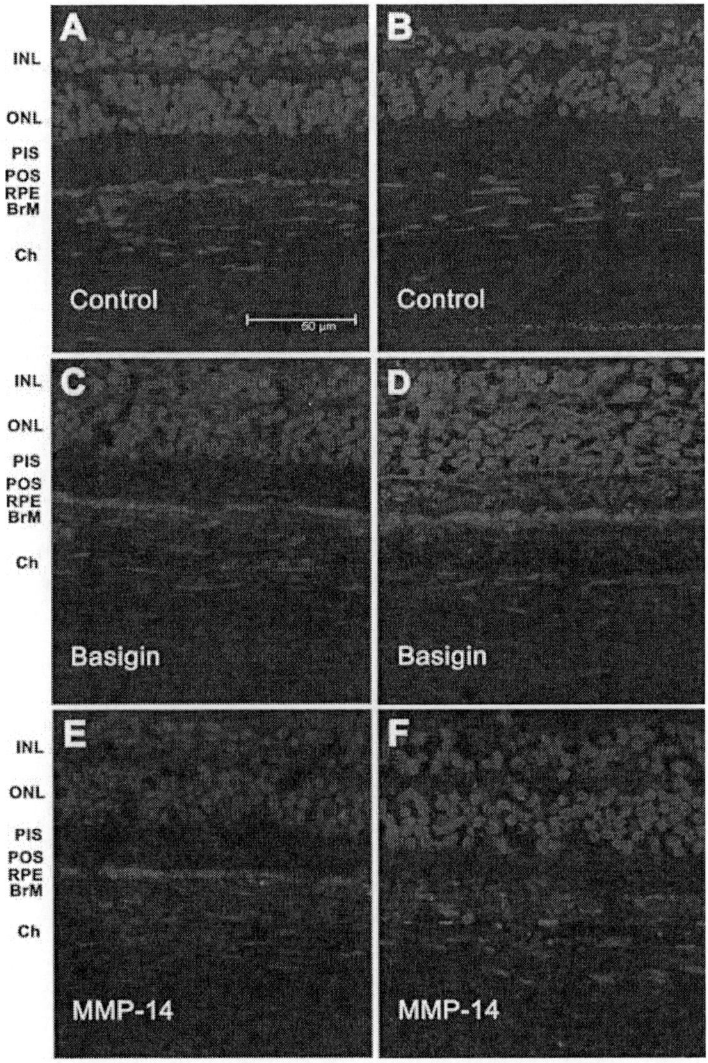

Figure 9. Immunohistochemical analysis of basigin and MMP-14 in human retina. Retina sections from human donor eyes with no known eye disease (A, C, and E) or from human donor eyes with dry AMD (B, C, and F) were stained with either mouse polyclonal anti-basigin (C and D) or mouse monoclonal anti-MMP-14 (E and F) as indicated. Negative controls were generated by omission of the primary antibody (A and B). Secondary antibodies were coupled to Alexa Fluor 488. Nuclei were stained with 4,6-diamidino-2-phenylindole dihydrochloride. Sections were analyzed under a confocal microscope. INL, inner nuclear layer; ONL, outer nuclear layer; PIS, photoreceptor inner segments; POS, photoreceptor outer segments; Ch, choroi.

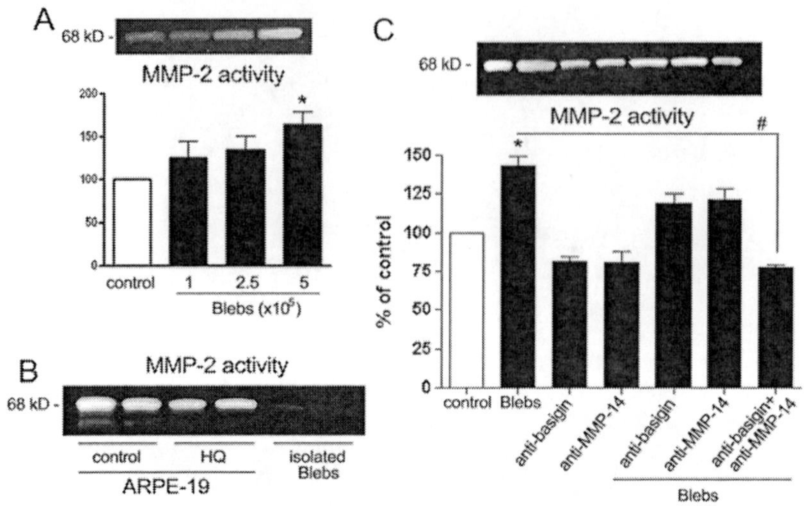

Figure 10. MMP-2 activity in ARPE-19 cells exposed to hydroquinone (HQ)-induced blebs. Blebs were generated by incubation with 100 µm HQ for 6 h and used to treat ARPE-19 cells for 24 h. Thereafter conditioned medium was collected, and MMP-2 activity was evaluated by gelatin zymography. *A*, MMP-2 activity from cells (4.5×10^5) incubated with increasing concentrations of blebs. B, MMP-2 activity in control, non-treated cells (C , HQ-treated cells (HQ), and isolated blebs alone (B). C, MMP-2 activity from cells incubated with blebs (5×10^5 blebs/4.5×10^5 cells) in the presence or absence of anti-basigin (1:500) and anti-MMP-14 (1:500) antibodies. *Top*, gelatin zymogram from a representative experiment. *Bottom*, bar graph showing averages of results of three independent experiments. Results are expressed as mean ± S.E. #, $p< 0.01$; *, $p< 0.05$ compared with control cells. Error bars indicate standard error of the mean.

In addition, our data showed that RPE cells incubated with blebs exhibit increased active MMP-2. MMP-2 activity returned to basal levels after incubation with anti-basigin and MMP-14 antibodies (Figure 10), suggesting that both proteins may play a pivotal role determining MMP-2 activity [192]. Thus, blebs accumulated under the RPE will stimulate ECM turnover increasing active MMP-2 through the action of two bleb-carried proteins, basigin and MMP-14. Therefore, we speculate that blebs may play an important role for sub-RPE to traverse RPE basement membrane. RPE are subsequently stimulated to increase synthesis of collagens and other molecules responsible for ECM turnover, affecting both RPE basement membrane and BrM. This process leads to the formation of new basement membrane under the RPE to trap these deposits within BrM. We postulate that various hormones and other plasma-derived molecules related to systemic health cofactors are implicated in this stage.

INFLAMMATION AND ANGIOGENESIS: ROLE OF HYDROQUINONE

Inflammation

Data from a number of laboratories provide compelling evidence that inflammatory and/or immune-mediated events may participate in the development of sub-RPE deposits formation and/or progression to CNV [50, 56, 220-222]. Based upon available data, a new paradigm has been introduced for sub-RPE formation and its relationship to AMD. This integrated hypothesis is based largely upon the dynamic interactions between those factors that induce and sustain chronic local inflammation at the level of the RPE-BrM-choroidal interface, and those mechanisms that attenuate it. Complement and immune complexes have been identified in drusen, but their pathogenic role has not been defined. This information has been recently reviewed [56-59]. Other investigators have observed that choroidal monocytes/ macrophages are present in human specimens of both early and late AMD [56, 57, 59, 69, 223, 224]. Macrophages have been detected along the choriocapillaris-side of BrM underlying areas of thick drusen or other deposits [48, 225-228] and processes from choroidal monocytes have been noted to insert into BrM deposits [225] Moreover, dentritic cells are often observed in the sub-RPE space in association with whole, or portions of, RPE cells that have been shunted into BrM, prior to the time that drusen are detectable. Therefore, macrophages and choroidal dentritic cells may be activated and recruited by locally damaged and/or sublethaly injured RPE cells. This idea is consistent with the data showing that macrophages and/or dentritic cells, and thus the innate immune system, can be activated by microenvironmental tissue damage [43, 229, 230]. However it remains to be determined whether drusen-associated macrophages and dentriric cells initiate a classical immune response involving T helper cells, secreted cytokines, elicit an inflammatory or complement-mediated response, or play some other role in the generation of drusen.

While it is largely recognized that macrophages accumulate in AMD lesions, there is ambiguity surrounding their role in the disease process with conflicting evidence regarding whether they might be helpful by scavenging accumulated debris and therefore protecting against CNV or harmful by stimulating CNV [119]. This might be due to the largely observational nature of human samples but also probably reflects different functions macrophages serve during distinct phases of the disease.

Based upon the assumption that injured RPE may serve as stimulatory factors that initiate macrophages recruitment and activation, it was determined if monocyte populations from individual AMD patients exhibited heterogeneity in terms of cytokine production in culture or in mRNA expression of freshly isolated cells, and if high levels served as a biomarker of risk for progression into neovascular AMD [70] TNF-α expression, a potent cytotoxic and proangiogenic cytokine for wet AMD. Typically human macrophages in culture, after stimulation with RPE derived blebs produced an increase in TNF-α. However, extremely high variability in baseline TNF-α expression from isolated macrophages was observed among different subjects [70]. These results are consistent with the hypothesis that the pre-existing macrophage activation state, defined as level of cytokine or mediator expression of the circulating monocyte, might determine the negative or positive consequence of macrophage recruitment as a disease modifier. Macrophages with low expression might remove deposits safely whereas macrophages with high expression might produce mediators that contribute to disease progression.

Aberrant expression of chemokines occurs in a variety of diseases that have an inflammatory component. Highly specialized RPE cells play a pivotal role in the maintenance of the outer retina by secreting several cytokines including monocyte chemoattractant protein-1 [MCP-1] [231, 232] which has been suggested to be implicated in the pathogenesis of AMD [233, 234]. RPE cells can secrete MCP-1 in the direction of choroidal blood vessels during inflammatory responses therefore suggesting that RPE cells might promote macrophage recruitment to the choroid from circulating monocytes [235]. It was reported that MCP-1 is regulated in injured ARPE-19 cells and that free radicals might be immunostimulatory [236] providing support for the notion that injured RPE cells may induce monocyte migration.

We have investigated MCP-1 expression in RPE from patients with AMD as well as the regulation of RPE-derived MCP-1 expression following cigarette hydroquinone-mediated oxidative injury. Our data report for the first time that MCP-1 expression is markedly decreased in RPE from smoker patients with AMD [237] thereby pointing to a critical role for MCP-1 in the pathogenesis of the disease. We acknowledge that due to the nature of our study, it cannot be determined whether the altered expression of MCP-1 in human RPE lysates is a cause or consequence of the disease. However, our current findings suggest that declining MCP-1 production by aging RPE cells may impair recruitment of macrophages/dentritic cells essential for scavenging debris which may lead to drusen formation and accumulation in AMD patients. It can be speculated that declining RPE-derived MCP-1 production resulting from cumulative exposure to oxidative damage may be an important factor that could accelerate and promote the formation of sub-RPE deposits in smokers. This theory is put forth bearing in mind the complexity of underlying cellular and molecular mechanisms involved in the inflammatory response and with the acknowledgment that numerous AMD genotypes may exist. Therefore, we fully recognize that only some aspects of the proposed hypothesis may be involved in any given AMD genotype.

Based on the above hypothesis, we have evaluated the possibility that HQ-induced oxidative stress might regulate MCP-1 expression in the RPE. We showed that prolonged exposure to hydroquinone-induced oxidative injury downregulated MCP-1 production by ARPE-19 cells and RPE/choroids from C57BL/6 mice [237]. An earlier study by Joly et al is in line with our observations showing a decline in MCP-1 gene expression in retinas of mice exposed to light-induced oxidative damage for several days [238]. Our observations suggest that sustained exposure to hydroquinone might impair RPE-derived MCP-1-mediated scavenging macrophages and dentritic cells recruitment and phagocytosis which might lead to incomplete clearance of proinflammatory debris trapped between the RPE and its BrM. On the other hand, our preliminary date show that RPE-derived blebs activate RPE MCP-1 production (unpublished data), suggesting that when significant BLD will have already formed, RPE-blebs and other debris will activate MCP-1 secretion by RPE leading to infiltration of proangiogenic macrophages to sub-RPE deposit areas, where they will scavenge, relate more cytokines and mediators, and amplify the process leading to progression of drusen to CNV in smoker patients with AMD.

Angiogenesis

Angiogenesis is a highly complex biological process that involves a delicate balance between numerous stimulators and inhibitors, each regulated by multiple control systems. CNV-related angiogenesis requires an alteration in the concentration of molecules that stimulate or inhibit growth of new blood vessels [239]. Vascular endothelial growth factor (VEGF), constitutively produced and secreted by RPE in culture [240-248], is a major angiogenic cytokine central to the development of wet AMD [242-245].VEGF regulates endothelial cells proliferation, migration and survival [249]. Interestingly, secretion of VEGF by RPE cells is polarized towards BrM [250]. There is ample clinical evidence that VEGF expression is increased in surgically excised AMD-associated choroidal neovascular membranes [245-247]. Eyes with early forms of AMD have increased expression of VEGF in the RPE and the vitreous of eyes with CNV have increased concentration of VEGF [245]. Similar observations have been made in animal models of CNV [241, 251]. Furthermore, Reich et al reported that subretinal injection of VEGF siRNA significantly inhibited the growth of laser-induced CNV in a mouse model [248]. PEDF, a potent angiogenic inhibitor [252] secreted by RPE cells [253-255], counterbalances the effects of VEGF and modulates the formation of CNV [253, 254]. A decrease in PEDF expression has been reported in eyes with AMD, therefore disrupting the critical balance between VEGF and PEDF that may lead to pathological angiogenesis and be permissive for the development of CNV [253]. PEDF levels decline in the vitreous of patients with CNV [255]. However, until now, there is no report in the literature examining VEGF and PEDF expression in RPE from AMD patients or evaluating whether or not cigarette smoke-related hydroquinone-induced oxidative stress has the potential to dysregulate the VEGF/PEDF balance in RPE cells.

Given their critical role in AMD and that oxidative damage to the RPE and angiogenesis appear to be central in the pathogenesis of the disease, we studied the expression of these chemokines in RPE from smoker patients diagnosed with AMD and whether hydroquinone-induced oxidative stress might also regulate VEGF and PEDF expression in RPE cells.

We report that VEGF expression is increased and PEDF expression is decreased in RPE from smoker patients with AMD resulting in an increased VEGF-to-PEDF ratio [237] A disruption in the critical balance of these opposing stimuli may be permissive for the development of wet AMD. Our findings are consistent with clinical observations describing dysregulated expression of VEGF and PEDF [245-247, 252, 253, 255] in eyes with AMD. Oxidant-mediated RPE damage might promote abnormal angiogenesis [239]. In vitro, although oxidative injury declined the production of PEDF without significantly changing VEGF expression in ARPE-19 cells regardless of dose and duration of exposure to hydroquinone, we observed an increased VEGF-to-PEDF ratio which may favor angiogenesis [237]. These results suggest that cigarette smoke-related HQ-induced oxidative stress might impair the delicate balance between VEGF and PEDF that controls angiogenic homeostasis in the retina. Other previous reports showed that cigarette smoke extract induces VEGF expression in ARPE-19 cells [124] and that H_2O_2-induced oxidative stress increased the production of VEGF in human RPE cells [239]. The discrepancy between our in vitro findings and those earlier observations with regards to VEGF might reflect differences in cellular responses in the setting of different types of oxidant-mediated injury.

The endogenous angiogenic inhibitors are believed to be essential for maintaining the homeostasis of angiogenesis in the retina. Given the evidence that PEDF is an important negative regulator of angiogenesis, lower levels of PEDF is strongly suggestive of a decreased anti-angiogenic activity that may lead to the initiation of angiogenesis in response to hydroquinone-induced oxidative stress. However, we do not rule out the possibility that a decreased level of inhibitory factor PEDF by itself may not be sufficient for inducing the angiogenic switch leading to CNV. Reciprocal increase in stimulatory VEGF might also be needed. In fact, a longer more sustained exposure to hydroquinone might be necessary to induce VEGF expression in ARPE-19 cells. Furthermore, angiogenesis is a highly complex and tightly orchestrated multistep process involving extensive interplay between multiple angiogenic factors. It is therefore possible that several other molecules besides VEGF and PEDF regulated by hydroquinone might permit the development of abnormal angiogenesis. In vivo, we observed elevated expression of VEGF and PEDF protein in RPE/choroids from HQ-treated mice which translated into an enhanced VEGF-to-PEDF ratio.

As stated earlier, important species-specific differences may also account for the discrepancy between human cells and mice results. In addition, one has to keep in mind that inherent in vitro and in vivo differences might explain this disparity. PEDF has multiple dose-dependent biological functions. Interestingly, it has been reported that low doses of PEDF are inhibitory but high doses can increase the development of CNV induced by laser in mice [256]. In addition, a study showed that RPE-derived VEGF upregulates PEDF expression via VEGF receptor-1 in an autocrine manner [257], therefore highlighting regulatory interactions between these two counterbalancing systems of angiogenic stimulators and inhibitors.

In any case, our in vivo findings confirm that hydroquinone-induced oxidative damage is unequivocally associated with an imbalance between VEGF and PEDF in the RPE.

Taken together, these data provide strong support for a key role played by hydroquinone-injured RPE cells in the progression of dry AMD to CNV. We demonstrated that RPE dysfunction might lead to dysregulation of macrophage clearance function and angiogenic homeostasis as a result of oxidative damage which may trigger progression towards CNV in smoker patients with dry AMD.

Hydroquinone and Animal Model for Dry AMD

Animal models are extremely useful in preclinical testing of theories of disease pathogenesis and can serve to question current hypotheses and predict outcomes of therapeutic interventions. Our laboratory and others have used the C57BL/6 mouse model to evaluate mechanisms for age, gender, diet, environmental toxins, and to text hypothesis [66, 126, 195, 196, 258].

Based on the idea that hydroquinone might contribute to drusen pathogenesis, we extended our in vitro data to a more physiological environment using the 16-month-old C57BL/6 mouse model for dry AMD published by our laboratory [62, 66, 195, 196], but providing an alternative source of oxidant stimulus by replacing exposure to blue light with exposure to hydroquinone in food [66] or drinking water [61]. In addition, mice received a regular fat diet instead of high-fat diet. We evaluated the impact of this compound on the development of sub-RPE deposits, by using TEM. As reported earlier, hydroquinone-treated mice had increased blood levels of hydroquinone relative to control mice that showed non detectable levels [61]. Mice not exposed to hydroquinone showed normal morphology of the RPE, BrM, and choriocapillaris endothelium (Figure 11A).

Some specimens demonstrated mild frequency of any BLD. None of the eyes in this group demonstrated moderate BLD. Animals exposed to hydroquinone showed pathologic changes in the RPE and BrM characterized by moderate BLD [61]. Approximately a 83% of eyes exhibited moderate BLD. The BrM was thickened, with coated vesicles, membranous profiles, and banded structures, (Figure 11B, 11C), typical of those described in some human AMD specimens [62]. Findings were of a magnitude similar to those previously observed in mice exposed to high-fat diet plus blue green light [62].

Interestingly, animals showed blebs (Figure 11C). In preliminary experiments, we have also demonstrated that mice receiving subconjunctival injections of hydroquinone exhibeted a rudimentary form of BLD, often demonstrating small vesicular vesicles bleblike structures (Reinoso, et al. IOVS 2005;46:ARVO E-Abstract 3010). In addition, RPE from mice exposed to hydroquinone in drinking water showed increased levels of phosphorylated Hsp25, p38 and ERK [197], suggesting that phosphorylated Hsp25 might be a key mediator in early cellular events associated with actin reorganization and bleb formation involved in sub-RPE deposits formation.

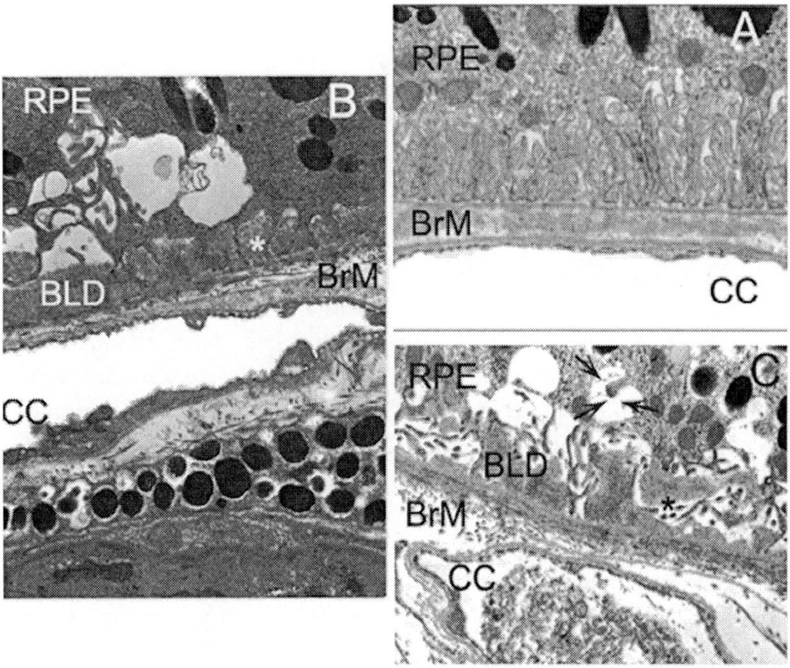

Figure 11. Transmission electron microscopy of the outer retina and choroid from a 16-month-old female mouse fed with a regular fat diet for 4 months. (**A**) Outer retina and choroid of a mouse fed a regular diet without oxidant showed a normal RPE, BrM, and choriocapillaris. (B) Outer retina and choroid of a mouse fed a regular diet and exposed to HQ (0.8% for 4 months) in drinking water revealed sub-RPE deposits characterized by accumulation of moderately severe BLD with dense granular material between the RPE and its basement membrane (*), compatible with a high mean severity score. The specimen, also, shows the abnormal choriocapillaris endothelium with increased thickening and loss of fenestration. (C) TEM of the outer retina and choroids from another 16-month-old female mouse fed a regular diet and exposed to HQ. The specimen shows moderately thick BLD with bandes structures (*) and occasional blebs (black arrows). CC, choriocapillaris, Magnification: (A, C) x25,000; (B) x7,200.

Taken together, these observations indicate that two different oxidant stimuli (i.e., blue light and hydroquinone) may induce a common response in the RPE and that a high-fat diet is not an absolute requirement for the development of BLD. Thus, the results suggest the role for blebs in sub-RPE formation and propose hydroquinone as another oxidative injury stimulus to the RPE, which may serve to explain the mechanisms that underlie pathologic BLD deposits in early AMD.

Conclusion

In summary, we postulate that cigarette smoke-related hydroquinone plays a role in the development of dry AMD and its progression to wet AMD. Although this hypothesis remains to be proven, we have proposed new ideas and suggested different mechanisms highlighting Hsp27, MMP-2, basigin, MMP-14, and MCP-1 as potential disease-related proteins as well as biochemical pathways for potential therapeutic strategies, which might result in prevention of more severe and irreversible late stages of this dreadful disease. Our goal is to intervene promptly in the early stages of the disease so that progression to the more severe late forms of AMD can be prevented. In this respect, RPE-derived MMP-2 activity and blebs are potential target due to their pivotal role in stimulating drusen formation and progression into CNV. Levels of phosphorylated Hsp27, glycosilated basigin and MMP-14, as well as, MCP-1 could be markers, which may contribute and aid to the ophthalmologic community in the management of the drusen. Drugs directed against Hsp27 phosphorylation pathway in addition to antioxidants could be used successfully on the prevention of sub-RPE deposits formation in selected high-risk AMD patiens.

Both the general population and many patients attending eye clinics are unaware of the strong link between smoking and AMD.The fear of going blind may be more motivational than warning about other tobacco-associate disease. Thus, efforts should be directed towards augmenting the campaign against tobacco smoking by adding the increased risk of blindness to the better-known arguments against smoking. We should urge our patients to quit smoking, and we must make them keenly aware of the afflictions that can develop when smoke gets in our body.

Grant support: NEI/NIH grants: EY13318, EY014477, and EY015249
Commercial relationship: None.

REFERENCES

[1] Shen, M; Lan, Q; Zhang, L; Chanock, S; Li, G; Vermeulen, R; Rappaport, SM; Guo, W; Hayes, RB; Linet, M; Yin, S; Yeager, M; Welch, R; Forrest, MS; Rothman, N; Smith MT. Polymorphisms in genes involved in DNA double-strand break repair pathway and susceptibility to benzene-induced hematotoxicity. *Carcinogenesis,* 2006;27(10):2083-2089.

[2] Shen, M; Zhang, L; Bonner, MR; Liu, CS; Li, G; Vermeulen, R; Dosemeci, M; Yin, S; Lan, Q. Association between mitochondrial DNA copy number, blood cell counts, and occupational benzene exposure. *Environ. Mol. Mutagen.*, 2008;49(6):453-457.

[3] Kondrová, E; Stopka, P; Soucek, P. Cytochrome P450 destruction by benzene metabolites 1,4-benzoquinone and 1,4-hydroquinone and the formation of hydroxyl radicals in minipig liver microsomes. *Toxicol. In Vitro,* 2007;21(4):566-75.

[4] Melikian, AA; Chen, KM; Li, H; Sodum, R; Fiala, E; El-Bayoumy, K. The role of nitric oxide on DNA damage induced by benzene metabolites. *Oncol. Rep.*, 2008;19(5):1331-1337.

[5] Ishiguro, S; Saugawara, H; Kusama, M; Yano, S; Shimojima, N; Sugawara, S. Glass capillary column gas chromatography analysis of tobacco and cellulose cigarette smoke. I. Acidic fraction. *Sci. Pap. Cent. Res. Inst.* Jpn, Tob. Salt Public Corp, 1976;118:207-221.

[6] Pryor, WA. Cigarette smoke radicals and the role of free radicals in chemical carcinogenicity. *Environ. Health Perspect.*, 1997; 105 (4): 875-882.

[7] Lenaz, G. Quinone specificity of complex I. *Biochim. Biophys. Acta,* 1998; 1364(2):207-221.

[8] Kari, FW; Bucher, J; Eustis, SL; Haseman, JK; Huff, JE. Toxicity and carcinogenicity of hydroquinone in F344/N rats and B6C3F1 mice. *Food Chem. Toxicol.,* 1992;30(9):737-747.

[9] Porteus, JW; Williams, RT. Studies in detoxication; the metabolism of benzene; the isolation of phenol, catechol, quinol and hydroxyquinol from ethereal sulphate fraction of the urine of rabbits receiving benzene orally. *Biochem. J.,* 1949;44(1):56-61.

[10] Parke, DV; Williams, RT. Studies in detoxication. 54. The metabolism of benzene. (a) The formation of phenylglucuronide and phenylsulphuric acid from (14C)benzene. (b)The metabolism of (14C)phenol. *Biochem. J.,* 1953;55(2):337-340.

[11] Monks, TJ; Lau, SS. Glutathione conjugation as a mechanism for the transport of reactive metabolites. *Adv. Pharmacol.*, 1994;27:183-210. Review.

[12] Winn, LM. Homologous recombination initiated by benzene metabolites: a potential role of oxidative stress. *Toxicol. Sci.*, 2003;72(1):143-149.

[13] Degli, EM. Inhibitors of NADH-ubiquinone reductase: an overview. *Biochim. Biophys. Acta,* 1998; 1364(2):222-235.

[14] McCue, JM; Link, KL; Eaton, SS; Freed, BM. Exposure to cigarette tar inhibits ribonucleotide reductase and blocks lymphocyte proliferation. *J. Immunol.*, 2000; 165(12):6771-6775.

[15] Gaido, KW; Wierda, D. Suppression of bone marrow stromal cell function by benzene and hydroquinone is ameliorated by indomethacin. *Toxicol. Appl. Pharmacol.*, 1987 ;89(3):378-390.

[16] Andreoli, C; Leopardi, P; Crebelli, R. Detection of DNA damage in human lymphocytes by alkaline single cell gel electrophoresis after exposure to benzene or benzene metabolites. *Mutat. Res.*, 1997;377(1):95-104.

[17] Ma, Q; Kinneer, K; Ye, J; Chen, BJ. Inhibition of nuclear factor kappaB by phenolic antioxidants: interplay between antioxidant signaling and inflammatory cytokine expression. *Mol. Pharmacol.,* 2003; 64(2): 211-219.

[18] Kim, AR; Cho, JY; Lee, JY; Choi, JS; Chung, HY. Hydroquinone modulates reactivity of peroxynitrite and nitric oxide production. *J. Pharm. Pharmacol.,* 2005;57(4):475-481.

[19] Lee, JY; Kim, JY; Lee, YG; Shin, WC; Chun, T; Rhee, MH; Cho, JY. Hydroquinone, a reactive metabolite of benzene, reduces macrophage-mediated immune responses. *Mol. Cells*, 2007;23(2):198-206.

[20] Haynes, WF Jr; Krstulovic, VJ; Bell, AL Jr. Smoking habit and incidence of respiratory tract infections in a group of adolescent males. *Am. Rev. Respir. Dis*., 1966;93(5):730-735.

[21] Nouri-Shirazi, M; Guinet, E. A possible mechanism linking cigarette smoke to higher incidence of respiratory infection and asthma. *Immunol. Lett.,* 2006;103(2):167-176.

[22] Cho, JY. Suppressive effect of hydroquinone, a benzene metabolite, on in vitro inflammatory responses mediated by macrophages, monocytes, and lymphocytes. Hindawi Publishing Corporation, *Mediators of inflammation,* 2008; Article ID 298010:1-11.

[23] Choi, JM; Cho, YC; Cho, WJ; Kim, TS; Kang, BY. Hydroquinone, a major component in cigarette smoke, reduces IFN-gamma production in antigen-primed lymphocytes. *Arch. Pharm. Res.*, 2008;31(3):337-341.

[24] Lee, MH; Chung, SW; Kang, BY; Kim, KM; Kim, TS. Hydroquinone, a reactive metabolite of benzene, enhances interleukin-4 production in CD4+ T cells and increases immunoglobulin E levels in antigen-primed mice. *Immunology,* 2002;106(4):496-502.

[25] Frazer-Abel, AA; McCue, JM; Lazis, S; Portas, M; Lambert, C; Freed, BM. Cigarette tar phenols impede T cell cycle progression by inhibiting cyclin-dependent kinases. *Mol. Immunol.*, 2007;44(4):488-493.

[26] Shen, DX; Shi, X; Fu, JL; Zhang, YM; Zhou, ZC.The role of thiol reduction in hydroquinone-induced apoptosis in HEK293 cells. *Chem. Biol. Interact.*, 2003;145(2):225-233.

[27] Scarpa, A; Guerci, A. Depigmenting procedures and drugs employed by melanoderm populations. *J. Ethnopharmacol.*, 1987 ;19(1):17-66. Review.

[28] Briganti, S; Camera, E; Picardo, M. Chemical and instrumental approaches to treat hyperpigmentation. *Pigment Cell Res., 2003*; 16(2):101-110. Review.

[29] Tse, TW. Hydroquinone for skin lightening: safety profile, duration of use and when should we stop? *J. Dermatolog. Treat.*, 2010;21(5):272-275. Review.

[30] Nissen, JN; Corydon, L. Corneal ulcer after exposure to vapours from bone cement (methyl methacrylate and hydroquinone). *Int. Arch. Occup. Environ. Health,* 1985;56(2):161-165.

[31] Gottfried, N. Corneal damage in hydroquinone workers. A clinico pathologic study. *Arch. Ophthal.*, 1996;76:189-194.

[32] Soucek, P. Cytochrome P450 destruction by quinones: comparison of effects in rat and human liver microsomes. *Chem. Biol. Interact.,* 1999;121(3):223-236.

[33] Gut, I; Nedelcheva, V; Soucek, P; Stopka, P; Tichavská, B. Cytochromes P450 in benzene metabolism and involvement of their metabolites and reactive oxygen species in toxicity. *Environ. Health Perspect.,* 1996;104 Suppl. 6:1211-1218.

[34] De Caprio, AP. The toxicology of hydroquinone: relevance to occupational and environmental exposure. *Crit. Rev. Toxicol.,* 1999;29(3):328-330.

[35] McDonald, TA; Holland, NT; Skibola, C; Duramand, P; Smith, MT. Hypothesis: phenol and hydroquinone derived mainly from diet and

gastrointestinal flora activity are causal factors in leukemia. *Leukemia,* 2001;15(1):10-20.

[36] Blacker, AM; Schroeder, RE; English, JC; Murphy, SJ; Krasavage, WJ; Simon, GS. A two-generation reproduction study with hydroquinone in rats. *Fundam. Appl. Toxicol.*, 1993;21(4):420-424.

[37] Fine, SL; Berger, JW; Maguire, MG; Ho. AC. Age-related macular degeneration. *N. Engl. J. Med.*, 2000; 342(7):483-492.

[38] Berger, JW; Fine, SL; Maguire, MG. *Age-related macular degeneration.* St. Louis: Mosby, 1999.

[39] Zarbin, MA. Current concepts in the pathogenesis of age-related macular degeneration. *Arch. Ophthalmol.*, 2004; 122(4):598-614.

[40] Abdelsalam, A; Del Priore, L; Zarbin, MA. Drusen in age-related macular degeneration: pathogenesis, natural course, and laser photocoagulation-induced regression. *Surv. Ophthalmol.*, 1999; 44(1): 1-29.

[41] Bird, AC; Bressler, NM; Bressler, SB; Chisholm, IH; Coscas, G; Davis, MD. An international classification and grading system for age-related maculopathy and age-related macular degeneration. The International ARM Epidemiological Study Group. *Surv. Ophthalmol.,* 1995; 39(5):367-374.

[42] Gass, JD. Drusen and disciform macular detachment and degeneration. *Trans Am. Ophthalmol. Soc.,* 1972; 70:409-436.

[43] Green, WR. Histopathology of age-related macular degeneration. *Mol. Vis.,* 1999; 5:27.:27.

[44] Curcio, CA; Medeiros, NE; Millican, CL. Photoreceptor loss in age-related macular degeneration. *Invest. Ophthalmol. Vis. Sci.*, 1996; 37(7):1236-1249.

[45] Abecasis, GR; Yashar, BM; Zhao, Y; Ghiasvand, NM; Zareparsi, S; Branham, KE. Age-related macular degeneration: a high-resolution genome scan for susceptibility loci in a population enriched for late-stage disease. *Am. J. Hum. Genet.*, 2004;74:482-494.

[46] Tuo, J; Bojanowski, CM; Chan, CC. Genetic factors of age-related macular degeneration. *Prog. Retin. Eye Res.*, 2004;23:229-249.

[47] Rodrigues, EB. Inflammation in dry age-related macular degeneration, *Ophthalmologica,* 2007;221;143-152.

[48] Oh, H; Takagi, H; Takagi, C. The potential angiogenic role of macrophages in the formation of choroidal neovascular membranes. *Invest. Ophthalmol. Vis. Sci.,* 1999;40:1891-1898.

[49] Liang, FQ; Godley, BF. Oxidative stress-induced mitochondrial DNA damage in human retinal pigment epithelial cells: a possible mechanism for RPE aging and age-related macular degeneration. *Exp. Eye Res.,* 2003;76:397-403.

[50] Donoso, LA; Kim, D; Frost, A; Callahan, A; Hageman, G. The role of inflammation in the pathogenesis of age-related macular degeneration. *Surv. Ophthalmol.*, 2006;51:137-152.

[51] Guymer, R; Luthert, P; Bird, A. Changes in Bruch's membrane and related structures with age. *Prog. Retin. Eye Res.,* 1999; 18(1):59-90.

[52] Inatani, M; Tanihara, H. Proteoglycans in retina. *Prog. Retin. Eye Res.,* 2002; 21(5):429-447.

[53] Hewitt, AT; Nakazawa, K; Newsome, DA. Analysis of newly synthesized Bruch's membrane proteoglycans. *Invest. Ophthalmol. Vis. Sci.,* 1989; 30(3):478-486.

[54] Young, RW. Pathophysiology of age-related macular degeneration. *Surv. Ophthalmol.*, 1987; 31[5]:291-306.

[55] Halliwell, B. Cigarette smoking and health: a radical view. *JR Soc. Health,* 1993;113:91–96.

[56] Hageman, GS; Luthert, PJ; Victor Chong, NH; Johnson, LV; Anderson, DH; Mullins, RF. An integrated hypothesis that considers drusen as biomarkers of immune-mediated processes at the RPE-Bruch's membrane interface in aging and age-related macular degeneration. *Prog. Retin. Eye Res.,* 2001; 20(6):705-732.

[57] Penfold, PL; Madigan, MC; Gillies, MC; Provis, JM. Immunological and aetiological aspects of macular degeneration. *Prog. Retin. Eye Res.,* 2001; 20(3):385-414.

[58] Hageman, GS; Mullins, RF. Molecular composition of drusen as related to substructural phenotype. *Mol. Vis.*, 1999;5:28. Review.

[59] Anderson, DH; Mullins, RF; Hageman, GS; Johnson, LV. A role for local inflammation in the formation of drusen in the aging eye. *Am. J. Ophthalmol.,* 2002; 134(3):411-431.

[60] Marin-Castaño, ME; Csaky, KG; Cousins, SW. Nonlethal oxidant injury to human retinal pigment epithelium cells causes cell membrane blebbing but decreased MMP-2 activity. *Invest. Ophthalmol. Vis. Sci.,* 2005;46(9):3331-3340.

[61] Marin-Castaño, ME; Striker, GE; Alcazar, O; Catanuto, P; Espinosa-Heidmann, DG; Cousins, SW. Repetitive nonlethal oxidant injury to retinal pigment epithelium decreased extracellular matrix turnover in

vitro and induced sub-RPE deposits in vivo. *Invest. Ophthalmol. Vis. Sci.,* 2006;47(9):4098-4112.

[62] Cousins, SW; Espinosa-Heidmann, DG; Alexandridou, A; Sall, J; Dubovy, S; Csaky, K. The role of aging, high fat diet and blue light exposure in an experimental mouse model for basal laminar deposit formation. *Exp. Eye Res.*, 2002; 75(5):543-553.

[63] Strunnikova, N; Zhang, C; Teichberg, D; Cousins, SW; Baffi J, Becker, KG; Csaky, KG. Survival of retinal pigment epithelium after exposure to prolonged oxidative injury: a detailed gene expression and cellular analysis. *Invest. Ophthalmol. Vis. Sci.,* 2004;45(10):3767-3777.

[64] Beatty, S; Koh, H; Phil, M; Henson, D; Boulton, M. The role of oxidative stress in the pathogenesis of age-related macular degeneration. *Surv. Ophthalmol.*, 2000; 45(2):115-134.

[65] Winkler, BS; Boulton, ME; Gottsch, JD; Sternberg, P. Oxidative damage and age-related macular degeneration. *Mol. Vis.*, 1999; 5:32.

[66] Espinosa-Heidmann, DG; Suner, IJ; Catanuto, P; Hernandez, EP; Marin-Castano, ME; Cousins, SW. Cigarette smoke-related oxidants and the development of sub-RPE deposits in an experimental animal model of dry AMD. *Invest. Ophthalmol. Vis. Sci.*, 2006;47(2):729-737.

[67] Zarbin, MA. Current concepts in the pathogenesis of age-related macular degeneration. *Arch. Ophthalmol.*, 2004; 122(4):598-614.

[68] Chisolm, GM III; Hazen, SL; Fox, PL; Cathcart, MK. The oxidation of lipoproteins by monocytes-macrophages. Biochemical and biological mechanisms. *J. Biol. Chem.,* 1999; 274(37):25959-25962.

[69] Grossniklaus, HE; Ling, JX; Wallace, TM; Dithmar, S; Lawson, DH; Cohen, C et al. Macrophage and retinal pigment epithelium expression of angiogenic cytokines in choroidal neovascularization. *Mol. Vis.,* 2002; 8:119-126.

[70] Cousins, SW; Espinosa-Heidmann, DG; Csaky, KG. Monocyte activation in patients with age-related macular degeneration: a biomarker of risk for choroidal neovascularization? *Arch. Ophthalmol.*, 2004; 122(7):1013-1018.

[71] Espinosa-Heidmann, DG; Suner, IJ; Hernandez, EP; Monroy, D; Csaky, KG; Cousins, SW. Macrophage depletion diminishes lesion size and severity in experimental choroidal neovascularization. *Invest. Ophthalmol. Vis. Sci.,* 2003; 44(8):3586-3592.

[72] Tong, JP; Yao, IF.Contribution of VEGF and PEDF to choroidal angiogenesis: a need for balanced expressions. *Clin. Biochem.,* 2006;39:267–276.

[73] McCarty, CA; Mukesh, BN; Fu, CL; Mitchell, P; Wang, JJ; Taylor, HR. Risk factors for age-related maculopathy: the Visual Impairment Project. *Arch. Ophthalmol.*, 2001; 119(10):1455-1462.

[74] Smith, W; Mitchell, P; Leeder, SR. Smoking and age-related maculopathy. The Blue Mountains Eye Study. *Arch. Ophthalmol.*, 1996; 114(12):1518-1523.

[75] Evans, JR. Risk factors for age-related macular degeneration. *Prog. Retin. Eye Res.*, 2001; 20(2):227-253.

[76] Ishibashi, T; Patterson, R; Ohnishi, Y; Inomata, H; Ryan, S. Formation of drusen in the human eye. *Am. J. Ophthalmol.*, 1986;101:342–353.

[77] Donders, F. Beitrage zur pathologischen Anatomie des Auges. Graefe's *Arch. Clin. Exp. Ophthalmol.*, 1854;1:106–118.

[78] Muller, H. Anatomische beitrage zur ophthalmologie. *Albrecht. von Graefe Arch. Ophthalmol.*, 1856; 2:1–69.

[79] Burns, R; Feeney Burns, L. Clinico-morphologic correlations of drusen of Bruch's membrane. *Trans Am. Ophthalmol. Soc.*, 1980;78:206–225.

[80] Farkas, T; Sylvester, V; Archer, D. The ultrastructure of drusen. *Am. J. Ophthalmol.*, 1971;71:1196–1205.

[81] Feeney-Burns, L; Gao, C; Tidwell, M.Lysosomal enzyme cytochemistry of human RPE, Bruch's membrane and drusen. *Invest. Ophthalmol. Visual Sci.*, 1987;28:1138–1147.

[82] El Baba, F; Green, W; Fleischmann, J; Finkelstein, D; de la Cruz, Z. Clinicopathologic correlation of lipidization and detachment of the retinal pigment epithelium. *Am. J. Ophthalmol.*, 1986;101:576–583.

[83] Esfandiary, H; Chakravarthy, U; Patterson, C; Young, I; Hughes, AE. Association study of detoxification genes in age related macular degeneration. *Br. J. Ophthalmol.*, 2005;89:470-474.

[84] Clemons, TE; Milton, RC; Klein, R; Seddon, JM; Ferris, FL. Age-Related Eye Disease Study Research Group. Risk factors for the incidence of Advanced Age-Related Macular Degeneration in the Age-Related Eye Disease Study (AREDS). *Ophthalmology*, 2005;112: 533-539.

[85] Seddon, JM; Willett, WC; Speizer, FE; Hankinson, SE. A prospective study of cigarette smoking and age-related macular degeneration in women. *JAMA*, 1996;276:1141-1146.

[86] Tomany, SC; Wang, JJ; Van Leeuwen, R; Klein, R; Mitchell, P; Vingerling, JR; Klein, BE; Smith,W; De Jong, PT. Risk factors for incident age-related macular degeneration: pooled findings from 3 continents. *Ophthalmology*, 2004;111:1280-1287.

[87] Ni Dhubhghaill, SS; Cahill, MT; Campbell, M; Cassidy, L; Humphries, MM; Humphries, P. The pathophysiology of cigarette smoking and age-related macular degeneration. *Adv. Exp. Med. Biol.*, 2010;664:437-446.

[88] Solberg, Y; Rosner, M; Belkin, M. The association between cigarette smoking and ocular diseases. *Surv. Ophthalmol.*, 1998;42:535–547.

[89] Rahman, I; MacNee, W. Role of oxidants/antioxidants in smoking-induced lung diseases. *Free Radical Biology and Medicine*, 1996;21(5):669–681.

[90] Liu,,Z; Sun, L; Zhu, L; Jia, X; Li, X; Jia, H; Wang, Y; Weber, P; Long, J; Liu, J. Hydroxytyrosol protects retinal pigment epithelial cells from acrolein-induced oxidative stress and mitochondrial dysfunction. *J. Neurochem.,* 2007;103(6):2690-2700.

[91] Feng, Z; Liu, Z; Li, X; Jia, H; Sun, L; Tian, C; Jia, L; Liu, J. Alpha-Tocopherol is an effective Phase II enzyme inducer: protective effects on acrolein-induced oxidative stress and mitochondrial dysfunction in human retinal pigment epithelial cells. *J. Nutr. Biochem.,* 2010;21[12]:1222-1231.

[92] Jia, L; Liu, Z; Sun, L; Miller, SS; Ames, BN; Cotman, CW; Liu, J. Acrolein, a toxicant in cigarette smoke, causes oxidative damage and mitochondrial dysfunction in RPE cells: protection by [R]-alpha-lipoic acid. *Invest. Ophthalmol. Vis. Sci.*, 2007;48(1):339-48.

[93] Takeuchi, A; Takeuchi, M; Oikawa, K; Sonoda, KH; Usui, Y; Okunuki, Y; Takeda, A; Oshima, Y; Yoshida, K; Usui, M; Goto, H; Kuroda, M. Effects of dioxin on vascular endothelial growth factor (VEGF) production in the retina associated with choroidal neovascularization. *Invest. Ophthalmol. Vis. Sci.*, 2009;50(7):3410-34166.

[94] Sharma, A; Neekhra, A; Gramajo, AL; Patil, J; Chwa, M; Kuppermann, BD; Kenney, MC. Effects of Benzo(e)Pyrene, a toxic component of cigarette smoke, on human retinal pigment epithelial cells in vitro. *Invest. Ophthalmol. Vis. Sci.,* 2008;49(11):5111-5117.

[95] Wills, NK; Kalariya, N; Sadagopa Ramanujam, VM; Lewis, JR; Haji Abdollahi, S; Husain, A; van Kuijk, FJ.Human retinal cadmium accumulation as a factor in the etiology of age-related macular degeneration. *Exp. Eye Re.*, 2009;89:79-87.

[96] Kalariya, NM; Wills, NK; Ramana, KV; Srivastava, SK; van Kuijk, FJ. Cadmium-induced apoptotic death of human retinal pigment epithelial cells is mediated by MAPK pathway. *Exp. Eye Res.*, 2009;89(4):494-502.

[97] Wills, NK; Sadagopa Ramanujam, VM; Chang-Strepka, J; Kalariya, N; Lewis, JR; van Kuijk, FJ. Cadmium accumulation in the aging human retina. *Exp. Eye Res.*, 2008;86(1):41-51.

[98] Halliwell, B. Free radicals and vascular disease: how much do we know? *BMJ,* 1993; 307(6909):885-886.

[99] Winston, GW; Church, DF; Cueto, R; Pryor, WA. Oxygen consumption and oxyradical production from microsomal reduction of aqueous extracts of cigarette tar. *Arch. Biochem. Biophys.*, 1993; 304(2):371-378.

[100] Niki, E; Minamisawa, S; Oikawa, M; Komuro, E. Membrane damage from lipid oxidation induced by free radicals and cigarette smoke. *Ann. N Y Acad. Sci.,* 1993; 686:29-37.

[101] Klein, R; Knudtson, MD; Cruickshanks, KJ; Klein, BE. Further observations on the association between smoking and the long-term incidence and progression of age-related macular degeneration: The Beaver Dam Eye Study. *Archives of Ophthalmology*, 2008;126(1): 115-121.

[102] Chow, CK; Thacker, RR; Chow, CK; Thacker, RR; Changchit, C; Bridges, RB; Rehm, SR; Humble, J; Turbek, J. Lower levels of vitamin C and carotenes in plasma of cigarette smokers. J. *Am. Coll. Nutr.,* 1986;5 (3):305-312.

[103] Church, DF; Pryor, WA. Free-radical chemistry of cigarette smoke and its toxicological implications. *Environ. Health Perspect*, 1985;64: 111-126.

[104] AREDS. A randomized, placebo-controlled, clinical trial of high-dose supplementation with vitamins C and E, beta carotene, and zinc for age-related macular degeneration and vision loss: AREDS report no. 8. *Archives of Ophthalmology*, 2001;119:1417-1436.

[105] Moriarty-Craige, SE; Adkison, J; Lynn, M; Gensler, G; Bressler, S; Jones, DP; Sternberg, P Jr. Antioxidant supplements prevent oxidation of cysteine/cystine redox in patients with age-related macular degeneration. *Am. J. Ophthalmol.*, 2005;140:1020-1026.

[106] Jakobsdottir, J; Conley, YP; Weeks, DE; Mah, TS; Ferrell, RE; Gorin, MB. Susceptibility genes for age-related maculopathy on chromosome 10q26. *Am. J. of Human Genetics*, 2005;77(3): 389-407.

[107] Rivera, A; Fisher, SA; Fritsche, LG; Keilhauer, CN; Lichtner, P; Meitinger, T; Weber, BH.Hypothetical LOC387715 is a second major susceptibility gene for age-related macular degeneration, contributing independently of complement factor H to disease risk. *Human Molecular Genetics,* 2005;14(21):3227-3236.

[108] Schmidt, S; Hauser, MA; Scott, WK; Postel, EA; Agarwal, A; Gallins, P; Wong, F; Chen, YS; Spencer, K; Schnetz-Boutaud, N; Haines, JL; Pericak-Vance, MA. Cigarette smoking strongly modifies the association of LOC387715 and age-related macular degeneration. *Am. J. of Human Genetics,* 2006;78(5): 852-864.

[109] Kanda, A; Chen, W; Othman, M; Branham, KE; Brooks, M; Khanna, R; He, S; Lyons, R; Abecasis, GR; Swaroop, A. A variant of mitochondrial protein LOC387715/ARMS2, not HTRA1, is strongly associated with age-related macular degeneration. *Proceedings of the National Academy of Sciences USA,* 2007;41:16227-16232.

[110] Gu, X; Meer, SG; Miyagi, M; Rayborn, ME; Hollyfield, JG; Crabb, JW; Salomon, RG. Carboxyethylpyrrole protein adducts and autoantibodies, biomarkers for age-related macular degeneration. *Jl of Biol. Chem.,* 2003;43:42027–42035.

[111] Sun, ML; Finnemann ,SC; Febbraio, M; Shan, L; Annangudi, SP; Podrez, EA; Hoppe, G; Darrow, R; Organisciak, DT; Salomon, RG; Silverstein, RL; Hazen, S. Light-induced oxidation of photoreceptor outer segment phospholipids generates ligands for CD36-mediated phagocytosis by retinal pigment epithelium: A potential mechanism for modulating outer segment phagocytosis under oxidant stress conditions. *J. Biol. Chem.,* 2006;281(7):4222-4230.

[112] Schutt, F; Bergmann, M; Holz, FG; Kopitz, J. Proteins modified by malondialdehyde, 4-hydroxynonenal, or advanced glycation end products in lipofuscin of human retinal pigment epithelium. *Invest. Ophthalmol. Vis. Sci.,* 2003;44(8):3663-3668.

[113] Schutt, F; Ueberle, B; Schnölzer, M; Holz, FG; Kopitz, J. Proteome analysis of lipofuscin in human retinal pigment epithelial cells. *FEBS Letters,* 2002;528:217-221.

[114] Crabb, JW; Miyagi, M; Gu, X; Shadrach, K; West, KA; Sakaguchi, H; Kamei, M; Hasan, A; Yan, L; Rayborn, ME; Salomon, RG; Hollyfield, JG.Drusen proteome analysis: An approach to the etiology of age-related macular degeneration. *Proc. Natl. Acad. Sci. U SA*, 2002;99(23):14682-14687.

[115] Farboud, B; Aotaki-Keen, A; Miyata, T; Hjelmeland, LM; Handa, JT. Development of a polyclonal antibody with broad epitope specificity for advanced glycation endproducts and localization of these epitopes in Bruch's membrane of the aging eye. *Molecular Vision*, 1999; 5:11.

[116] Biswas, SK; Rahman, I. Environmental toxicity, redox signaling and lung inflammation: The role of glutathione. *Molecular Aspects of Medicine,* 2009;30(1-2):60-76.

[117] Edwards, AO; Ritter, R 3rd; Abel, KJ; Manning, A; Panhuysen, C; Farrer, LA.Complement factor H polymorphism and age-related macular degeneration. *Science*, 2005;308(5720):421-424.

[118] Zareparsi, S; Branham, KE; Li, M; Shah, S; Klein, RJ; Ott, J; Hoh, J; Abecasis, GR; Swaroop,A.Strong association of the Y402H variant in complement factor H at 1q32 with susceptibility to age-related macular degeneration. *Am. J. of Hum. Gen.*, 2005;77(1):149-153.

[119] Ding, X; Patel, M; Chan, CC. Molecular pathology of age-related macular degeneration. *Prog. Retin. Eye Res.*, 2009 ;28(1):1-18.

[120] Chen, H; Liu, B; Lukas, TJ; Neufeld, AH. The aged retinal pigment epithelium/choroid: a potential substratum for the pathogenesis of age-related macular degeneration *PLoS One*, 2008 Jun 4;3(6):e2339.

[121] Scott, WK; Schmidt, S; Hauser, MA; Gallins, P; Schnetz-Boutaud, N; Spencer, KL; Gilbert, JR; Agarwal, A; Postel, EA; Haines, Jl; Pericak-Vance, MA. Independent effects of complement factor H Y402H polymorphism and cigarette smoking on risk of age-related macular degeneration. *Ophthalmology,* 2007;114:1151-1156.

[122] Craig ,WY; Palomaki, GE; Haddow,JE. Cigarette smoking and serum lipid and lipoprotein concentrations: an analysis of published data. *BMJ,* 1989;298:784-788.

[123] Schmidt, S; Haines, JL; Postel, EA; Agarwal, A; Kwan, SY; Gilbert, JR; Pericak-Vance, MA; Scott, WK. Joint effects of smoking history and APOE genotypes in age-related macular degeneration. *Mol. Vis.,* 2005;11:941-949.

[124] Bertram, KM; Baglole, CJ; Phipps, RP; Libby, RT. Molecular regulation of cigarette smoke induced-oxidative stress in human retinal pigment epithelial cells: implications for age-related macular degeneration. *Am. J. Physiol. Cell Physiol.,* 2009;297(5):C1200-1210.

[125] Wang, AL; Lukas, TJ; Yuan, M; Neufeld, AH. Increased mitochondrial DNA damage and down-regulation of DNA repair enzymes in aged rodent retinal pigment epithelium and choroid. *Mol. Vis.,* 2008;(4):14:644-51.

[126] Wang, AL; Lukas, TJ; Yuan, M; Du, N; Handa, JT; Neufeld, AH. Changes in retinal pigment epithelium related to cigarette smoke: possible relevance to smoking as a risk factor for age-related macular degeneration. *PLoS One*, 2009;4(4):e5304.

[127] Hollyfield, JG; Bonilha, VL; Rayborn, ME; Yang, X; Shadrach, KG; Lu, L; Ugret, RL; Salomon, RG; Perez, VL. Oxidative damage-induced inflammation initiates age related macular degeneration. *Nat. Med.*, 2008:14(2):194-198.

[128] Mahapatra, SK; Das, S; Bhattacharjee, S; Gautam, N; Majumdar, S; Roy, S. *Toxicol. Mech. Methods*, 2009;19(2):100-108.

[129] Maritz, GS. Nicotine and lung development. *Birth Defects Res. C Embryo Today*, 2008;84(1):45-53.

[130] Das, S; Neogy, S; Gautan, N; Roy, S. In vitro nicotine induced superoxide mediated DNA fragmentation in lymphocytes: protective role of Andrographis paniculata Nees. *Toxicology in Vitro*, 2009;23: 909-998.

[131] Vassallo, R; Kroening, PR; Parambil, J; Kita, H. Nicotine and cigarette smoke constituents induce-immune-modulatory and pro-inflammatory dendritic cell responses. *Mol. Immunol.*, 2008;45(12):3321-3329.

[132] Pestana, IA; Vazquez-Padron, RI; Aitouche, A; Pham, SM. Nicotine and PDGF-receptor function are essential for nicotine-stimulated mitogenesis in human vascular smooth muscle cells. *J. Cell Biochem.*, 2005;965:986-995.

[133] Suner, I; Espinosa-Heidmann, DG; Marin-Castano, ME; Hernandez, E; Pereira-Simon, S; Cousins, SW. Nicotine increase size and severity of experimental choroidal neovascularization. *Invest. Ophthalmol. Vis. Sci.*, 2004;451:311-317.

[134] Villablanca, AC. Nicotine stimulates DNA synthesis and proliferation in vascular endothelial cell in vitro. *J. Appl. Physiol.*, 1998;846:2086-2098.

[135] Kiuchi, K; Matsuoka, M; Wu, JC; Lima e Silva, R; Kengatharan, M; Verghese, M; Ueno, S; Yokoi, K; Khu, NH; Cooke, JP; Campochiaro, PA. Mecamylamine suppresses basal and nicotine-stimulated choroidal neovascularization. *Invest. Ophthalmol. Vis. Sci.*, 2008 ;49(4):1705-11.

[136] Kanda, Y; Watanabe, Y. Nicotine-induced vascular endothelial growth factor release via the EGFR-ERK pathway in rat vascular smooth muscle cells. *Life Sci.*, 2007;8015:1409-1414.

[137] Ishibashi, T; Hata, Y; Yosikawa, H; Nakagawa, K; Sueishi, K; Inomata, H. Expression of vascular endothelial growth factor in experimental choroidal neovascularization. *Graefes Arch. Clin. Exp. Ophthalmol.*, 1997;2353: 159-167.

[138] Kliffen, M; Sharma, HS; Mooy, CM; Kerkvliet, S; de Jong, PT. Increased expression of angiogenic growth factor in age-related maculopathy. *Br. J. Ophthalmology*, 1997;812:154-162.

[139] Kvanta, A. Ocular angiogenesis: the role of growth factors. *Acta Ophthalmol.,* 2006. 843:282-288.

[140] Eisner, MD; Blanc, PD. Environmental tobacco smoke exposure during travel among adults with asthma. *Chest*, 2002;122:826–828.

[141] Jin, GF; Hurst, JS; Godley, BF. Hydrogen peroxide stimulates apoptosis in cultured human retinal pigment epithelial cells. *Curr. Eye Res.,* 2001;22:165–173.

[142] Yamada, H; Yamada, E; Hackett, SF; Ozaki, H; Okamoto, N; Campochiaro, PA. Hyperoxia causes decreased expression of vascular endothelial growth factor and endothelial cell apoptosis in adult retina. *J. Cell Physiol.*, 1999;179:149–156.

[143] Hecquet, C; Lefevre, G; Valtink, M; Engelmann, K; Mascarelli, F. Activation and role of MAP kinase-dependent pathways in retinal pigment epithelium cells: JNK1, P38 kinase, and cell death. *Invest. Ophthalmol. Vis. Sci.,* 2003;44:1320–1329.

[144] Dunaief, JL; Dentchev, T; Ying, GS; Milam, AH. The role of apoptosis in age-related macular degeneration. *Arch. Ophthalmol.,* 2002;120:1435–1442.

[145] Malorni, W; Donelli, G. Cell death: general features and morphological aspects. *Ann. NY Acad. Sci.,* 1992;663:218–233.

[146] Malorni, W; Donelli, G; Straface, E; Santini, MT; Paradisi, S; Giacomoni, PU. Both UVA and UVB induce cytoskeleton-dependent surface blebbing in epidermoid cells. *J. Photochem. Photobiol.,* 1994;26:265–270.

[147] van der Schaft, TL; de Bruijn, BC; Mooy, GM; Ketelaars, DAM; Jong, PTVM. Is basal laminar deposit unique for age-related macular degeneration?. *Arch. Ophthalmol.*, 1991;109:420–425.

[148] Ishibashi, T; Sorgente, N; Patterson, R; Ryan, SJ. Pathogenesis of drusen in the primate. *Invest. Ophthalmol. Vis. Sci.,* 1986;27:184-193.

[149] Zhu, ZR; Goodnight, R; Nishimura, T; Sorgente, N; Ogden, TE; Ryan, SJ. Experimental changes resembling the pathology of drusen in Bruch's membrane in the rabbit. *Curr. Eye Res,* 1988;7:581-592.

[150] Malorni, W; Iosi, F; Mirabelli, F; Bellomo, G. Cytoskeleton as a target in menadione-induced oxidative stress in cultured mammalian cells: alterations underlying surface bleb formation. *Chem. Biol. Interact.,* 1991;80:217-236.

[151] Gerthoffer, WT; Gunst, SJ. Invited review: focal adhesion and small heat shock proteins in the regulation of actin remodeling and contractility in smooth muscle. *J. Appl. Physiol.*, 2001;91:963-972.

[152] Dalle-Donne, I; Rossi, R; Milzani, A; Di Simplicio, P; Colombo, R. The actin cytoskeleton response to oxidants: from small heat shock protein phosphorylation to changes in the redox state of actin itself. *Free Radic. Biol. Med.,* 2001; 31(12):1624-1632.

[153] Lanzetti, L; Di Fiore, PP; Scita, G. Pathways linking endocytosis and actin cytoskeleton in mammalian cells. *Exp. Cell Res.,* 2001; 271(1): 45-56.

[154] Taunton, J. Actin filament nucleation by endosomes, lysosomes and secretory vesicles. *Curr. Opin. Cell Biol.,* 2001; 13(1):85-91.

[155] Rao, A; Craig, AM. Signaling between the actin cytoskeleton and the postsynaptic density of dendritic spines. *Hippocampus,* 2000; 10(5): 527-541.

[156] Kaibuchi, K; Kuroda, S; Amano, M. Regulation of the cytoskeleton and cell adhesion by the Rho family GTPases in mammalian cells. *Annu. Rev. Biochem.,* 1999; 68:459-486.

[157] Valentijn, K; Valentijn, JA; Jamieson, JD. Role of actin in regulated exocytosis and compensatory membrane retrieval: insights from an old acquaintance. *Biochem. Biophys. Res. Commun.,* 1999;266(3):652-661.

[158] Downey, GP. Mechanisms of leukocyte motility and chemotaxis. *Curr. Opin. Immunol.,* 1994;6(1):113-124.

[159] Hirokawa, N. Axonal transport and the cytoskeleton. *Curr. Opin. Neurobiol.,* 1993; 3(5):724-731.

[160] Weed, SA; Parsons, JT. Cortactin: coupling membrane dynamics to cortical actin assembly. *Oncogene,* 2001; 20(44):6418-6434.

[161] Milzani, A; DalleDonne, I; Colombo, R; Rodriguez OC. Prolonged oxidative stress on actin. *Arch. Biochem. Biophys.,* 1997;339(2): 267-274.

[162] Huot, J; Houle, F; Rousseau, S; Deschesnes, RG; Shah, GM; Landry, J. SAPK2/p38-dependent F-actin reorganization regulates early membrane blebbing during stress-induced apoptosis. *J. Cell Biol.,* 1998; 143 (5): 1361-1373.

[163] Concannon, CG; Gorman, AM; Samali, A. On the role of Hsp27 in regulating apoptosis. *Apoptosis.* 2003;8(1):61-70.

[164] Delogu, G; Signore, M; Mechelli, A; Famularo, G. Heat shock proteins and their role in heart injury. *Curr. Opin. Crit. Care,* 2002;8(5):411-416

[165] Haslbeck, M. sHsps and their role in the chaperone network. C*ell Mol. Life Sci.,* 2002;59(10):1649-1657.

[166] Constantin, B; Meerschaert, K; Vandekerckhove, J; Gettemans, J. Disruption of the actin cytoskeleton of mammalian cells by the capping

complex actin-fragmin is inhibited by actin phosphorylation and regulated by Ca2+ ions. *J. Cell Sci.*, 1998;111(Pt 12):1695-1706.

[167] Schneider, GB; Hamano, H; Cooper, LF. In vivo evaluation of hsp27 as an inhibitor of actin polymerization: hsp27 limits actin stress fiber and focal adhesion formation after heat shock. *J. Cell Physiol.,* 1998;177 (4):575-584.

[168] Ganea, E. Chaperone-like activity of alpha-crystallin and other small heat shock proteins. *Curr. Protein Pept. Sci.* 2001;2(3):205-225.

[169] de Lanerolle, P; Cole, AB. Cytoskeletal proteins and gene regulation: form, function, and signal transduction in the nucleus. *Sci. STKE,* 2002;2002 (139):E30.

[170] Landry, J; Huot, J. Modulation of actin dynamics during stress and physiological stimulation by a signaling pathway involving p38 MAP kinase and heat-shock protein 27. *Biochem. Cell Biol.*, 1995; 73 (9-10):703-707.

[171] Stokoe, D; Engel, K; Campbell, DG; Cohen, P; Gaestel, M. Identification of MAPKAP kinase 2 as a major enzyme responsible for the phosphorylation of the small mammalian heat shock proteins. *FEBS Lett.,* 1992;313(3):307-313.

[172] Huot, J; Houle, F; Marceau, F; Landry, J. Oxidative stress-induced actin reorganization mediated by the p38 mitogen-activated protein kinase/heat shock protein 27 pathway in vascular endothelial cells. *Circ. Res.,* 1997;80:383-392.

[173] Lavoie, JN; Lambert, H; Hickey, E; Weber, LA; Landry, J. Modulation of cellular thermoresistance and actin filament stability accompanies phosphorylation-induced changes in the oligomeric structure of heat shock protein 27. *Mol. Cell Biol.*, 1995;15:505-516.

[174] Guay, J; Lambert, H; Gingras-Breton, G; Lavoie, JN; Huot, J; Landry, J. Regulation of actinfilament dynamics by p38 map kinase-mediated phosphorylation of heat shock protein 27. *J. Cell Sci.*, 1997;110(3): 357-368.

[175] Clerk, A; Michael, A; Sugden, PH. Stimulation of multiple mitogen-activated protein kinase sub-families by oxidative stress and phosphorylation of the small heat shock protein, HSP25/27, in neonatal ventricular myocytes. *Biochem. J.,* 1998;333(3):581-589.

[176] Gusev, NB; Bogatcheva, NV; Marston, SB. Structure and properties of small heat shock proteins (sHsp) and their interaction with cytoskeleton proteins. *Biochemistry (Mosc.),* 2002;67:511-519.

[177] Shi, B; Han, B; Schwab, IR; Isseroff, RR. UVB irradiation-induced changes in the 27-kd heat shock protein (HSP27) in human corneal epithelial cells. *Cornea*, 2006;25:948-955.

[178] Decanini, A; Nordgaard, CL; Feng, X; Ferrington, DA; Olsen, TW. Changes in select redox proteins of the retinal pigment epithelium in age-related macular degeneration. *Am. J. Ophthalmol.*, 2007;143: 607-615.

[179] Pons, M; Cousins, SW; Csaky, KG; Striker, G; Marin-Castaño, ME. Cigarette smoke-related hydroquinone induces filamentous actin reorganization and heat shock protein 27 phosphorylation through p38 and extracellular signal-regulated kinase 1/2 in retinal pigment epithelium: implications for age-related macular degeneration. *Am. J. Pathol.,* 2010;177(3):1198-213.

[180] Strunnikova, N; Baffi, J; Gonzalez, A; Silk, W; Cousins, SW; Csaky, KG. Regulated heat shock protein 27 expression in human retinal pigment epithelium. *Invest. Ophthalmol. Vis. Sci.*, 2001;42:2130-2138.

[181] Cuenda, A; Rouse, J; Doza, YN; Meier, R; Cohen, P; Gallagher, TF; Young, PR; Lee, JC. SB203580 is a specific inhibitor of a MAP kinase homologue which is stimulated by cellular stresses and interleukin-1. *FEBS Lett.,* 1995;364:229-233.

[182] Huot, J; Lambert, H; Lavoie, JN; Guimond, A; Houle, F; Landry, J. Characterization of 45-kDa/54-kDa HSP27 kinase, a stress-sensitive kinase which may activate the phosphorylationdependent protective function of mammalian 27-kDa heat-shock protein HSP27. *Eur. J. Biochem.,* 1995;227:416-427.

[183] Cairns, J; Qin, S; Philp, R; Tan, YH; Guy, GR. Dephosphorylation of the small heat shock protein Hsp27 in vivo by protein phosphatase 2A. *J. Biol. Chem.,* 1994; 269:9176-9183.

[184] Kostenko S; Moens, U. Heat shock protein 27 phosphorylation: kinases, phosphatases,functions and pathology. *Cell Mol. Life Sci.,* 2009;66:3289-3307.

[185] Garg, TK; Chang, JY. Oxidative stress causes ERK phosphorylation and cell death in cultured retinal pigment epithelium: prevention of cell death by AG126 and 15-deoxy-delta 12,14-PGJ2. *BMC Ophthalmol.*, 2003; 3:5.

[186] Aumailley, M; Gayraud, B. Structure and biological activity of the extracellular matrix. *J. Mol. Med.*,1998;76:253-265.

[187] Kugler, A. Matrix metalloproteinases and their inhibitors. *Anticancer Res.,* 1999;19:1589-1592.

[188] Jacot, TA; Striker, GE; Stetler-Stevenson, M. Mesangial cells from transgenic mice with progressive glomerulosclerosis exhibit stable, phenotypic changes including undetectable MMP-9 and increased type IV collagen. *Lab. Invest.*, 1996;75:791-799.

[189] Peten, EP; Garcia-Perez, A; Terada, Y. Age-related changes in alpha 1- and alpha 2-chain type IV collagen mRNAs in adult mouse glomeruli: competitive PCR. *Am. J. Physiol.*, 1992;263:F951-F957.

[190] Belkhiri, A; Richards, C; Whaley, M. Increased expression of activated matrix metalloproteinase-2 by human endothelial cells after sublethal H2O2 exposure. *Lab. Invest.*, 1997;77:533-539.

[191] Alcazar, O; Cousins, SW; Marin-Castaño, ME. MMP-14 and TIMP-2 overexpression protects against hydroquinone-induced oxidant injury in RPE: implications for extracellular matrix turnover. *Invest. Ophthalmol. Vis. Sci.,* 2007;48(12):5662-70.

[192] Alcazar, O; Hawkridge, AM; Collier, TS; Cousins, SW; Bhattacharya, SK; Muddiman, DC; Marin-Castano, ME. Proteomics characterization of cell membrane blebs in human retinal pigment epithelium cells. *Mol. Cell Proteomics.,* 2009;8(10):2201-2211.

[193] Wiesner, C; Faix, J; Himmel, M; Bentzien, F; Linder, S. KIF5B and KIF3A/KIF3B kinesins drive MT1-MMP surface exposure, CD44 shedding, and extracellular matrix degradation in primary macrophages. *Blood,* 2010;116(9):1559-69.

[194] Zhang, C; Baffi, J; Cousins, SW; Csaky, KG. Oxidant-induced cell death in retinal pigment epithelium cells mediated through the release of apoptosis-inducing factor. *J. Cell Sci.,* 2003;116(10):1915-1923.

[195] Espinosa-Heidmann, DG; Sall, J; Hernandez, EP; Cousins, SW. Basal laminar deposit formation in APO B100 transgenic mice: complex interactions between dietary fat, blue light, and vitamin E. *Invest. Ophthalmol. Vis. Sci.*, 2004;45(1):260-266.

[196] Cousins, SW; Marin-Castano, ME; Espinosa-Heidmann, DG; Alexandridou, A; Striker, L; Elliot, S. Female gender, estrogen loss, and Sub-RPE deposit formation in aged mice. *Invest. Ophthalmol. Vis. Sci.,* 2003;44(3):1221-1229.

[197] Fossum, S; Mallett, S; Barclay, AN. The MRC OX-47 antigen is a member of the immunoglobulin superfamily with an unusual transmembrane sequence. *Eur.J. Immunol.,* 1991;21:671-679.

[198] Miyauchi, T; Masuzawa, Y; Muramatsu, T. The basigin group of the immunoglobulin superfamily: complete conservation of a segment in

and around transmembrane domains of human and mouse basigin and chicken HT7 antigen. *J. Biochem.*, 1991;110: 770-774.

[199] Kasinrerk, W; Fiebiger, E; Stefanová, I; Baumruker, T; Knapp, W; Stockinger, H. Human leukocyte activation antigen M6, a member of the Ig superfamily, is the species homologue of rat OX-47, mouse basigin, and chicken HT7 molecule. *J. Immunol.*, 1992;149: 847-854.

[200] Kasinrerk, W; Tokrasinwit, N; Phunpae, P. CD147 monoclonal antibodies induce homotypic cell aggregation of monocytic cell line U937 via LFA-1/ICAM-1 pathway. *Immunology*, 1999;96:184-192.

[201] Biswas, C; Zhang, Y; DeCastro, R; Guo, H; Nakamura, T; Kataoka, H; Nabeshima, K. The human tumor cell-derived collagenase stimulatory factor (renamed EMMPRIN) is a member of the immunoglobulin superfamily. *Cancer Res.*, 1995; 55: 434-439.

[202] Biswas, C. Tumor cell stimulation of collagenase production by fibroblasts. *Biochem. Biophys. Res. Commun.*, 1982;109:1026-1034.

[203] Kataoka, H; DeCastro, R; Zucker, S; Biswas, C. Tumor cell-derived collagenase-stimulatory factor increases expression of interstitial collagenase, stromelysin, and 72-kDa gelatinase. *Cancer Res.*, 1993; 53:3154-3148.

[204] Guo, H; Zucker, S; Gordon, MK; Toole, BP; Biswas, C. Stimulation of matrix metalloproteinase production by recombinant extracellular matrix metalloproteinase inducer from transfected Chinese hamster ovary cells. *J. Biol. Chem.*, 1997;272: 24-27.

[205] Sameshima, T; Nabeshima, K; Toole, BP; Yokogami, K; Okada, Y; Goya, T; Koono, M; Wakisaka, S. Glioma cell extracellular matrix metalloproteinase inducer (EMMPRIN) (CD147)stimulates production of membrane-type matrix metalloproteinases and activated gelatinase A in co-cultures with brain-derived fibroblasts. *Cancer Lett.*, 2000;157:177-184.

[206] Hakomori, S. Tumor malignancy defined by aberrant glycosylation and sphingo[glyco]lipid metabolism. *Cancer Res.*, 1996;56:5309-5318.

[207] Sun, J; Hemler, ME. Regulation of MMP-1 and MMP-2 production through CD147/extracellular matrix metalloproteinase inducer interactions. *Cancer Res.*, 2001;61:2276-2281.

[208] Tang, W; Chang, SB; Hemler, ME. Links between CD147 function, glycosylation, and caveolin-1. *Mol. Biol. Cell*, 2004;15:4043-4050.

[209] Ochrietor, JD; Moroz, TP; Clamp, MF; Timmers, AM; Muramatsu, T; Linser, PJ. Inactivation of the Basigin gene impairs normal retinal development and maturation. *Vision Res.*, 2002;42:447-453.

[210] Ochrietor, JD; Moroz, TP; van Ekeris, L; Clamp, MF; Jefferson, SC; deCarvalho, AC; Fadool, JM; Wistow, G; Muramatsu, T; Linser, PJ. Retina-specific expression of 5A11/Basigin-2, a member of the immunoglobulin gene superfamily. *Invest. Ophthalmol. Vis. Sci.,* 2003;44:4086-4096.

[211] Hori, K; Katayama, N; Kachi, S; Kondo, M; Kadomatsu, K; Usukura, J; Muramatsu, T; Mori, S; Miyake, Y. Retinal dysfunction in basigin deficiency. *Invest. Ophthalmol. Vis. Sci.*, 2000;41:3128-3133.

[212] Ochrietor, JD; Moroz, TM; Kadomatsu, K; Muramatsu, T; Linser, PJ. Retinal degeneration following failed photoreceptor maturation in 5A11/basigin null mice. *Exp. Eye Res.*, 2001;72:467-477.

[213] Clamp, MF; Ochrietor, JD; Moroz, TP; Linser, PJ. Developmental analyses of 5A11/Basigin, 5A11/Basigin-2 and their putative binding partner MCT1 in the mouse eye. *Exp. Eye Res.,* 2004;78:777-789.

[214] Taylor, PM; Woodfield, RJ; Hodgkin, MN; Pettitt, TR; Martin, A; Kerr, DJ; Wakelam, MJ. Breast cancer cell-derived EMMPRIN stimulates fibroblast MMP2 release through a phospholipase A[2] and 5-lipoxygenase catalyzed pathway. *Oncogene*, 2002; 21:5765-5772.

[215] Lim, M; Martinez, T; Jablons, D; Cameron, R; Guo, H; Toole, B; Li, JD; Basbaum, C. Tumor-derived EMMPRIN (extracellular matrix metalloproteinase inducer) stimulates collagenase transcription through MAPK p38. *FEBS Lett.*, 1998;441:88-92.

[216] Sidhu, SS; Mengistab, AT; Tauscher, AN; LaVail, J; Basbaum, C. The microvesicle as a vehicle for EMMPRIN in tumor-stromal interactions. *Oncogene.* 2004; 23:956-563.

[217] Seiki, M; Koshikawa, N; Yana, I. Role of pericellular proteolysis by membrane-type 1 matrix metalloproteinase in cancer invasion and angiogenesis. *Cancer Metastasis Rev.*, 2003;22:129-143.

[218] Itoh, Y; Seiki, M. MT1-MMP: a potent modifier of pericellular microenvironment. *J. Cell Physiol.*, 2006;206:1-8.

[219] Egawa, N; Koshikawa, N; Tomari, T; Nabeshima, K; Isobe, T; Seiki, M. Membrane type 1 matrix metalloproteinase (MT1-MMP/MMP-14) cleaves and releases a 22-kDa extracellular matrix metalloproteinase inducer (EMMPRIN) fragment from tumor cells. *J. Biol. Chem.,* 2006; 281:37576-37585.

[220] Patel, M; Chan, CC. Immunopathological aspects of age-related macular degeneration. *Semin. Immunopathol.*, 2008, 30:97-110.

[221] Haines, JL; Hauser, MA; Schmidt, S; Scott, WK; Olson, LM; Gallins, P; Spencer, KL; Kwan, SY; Noureddine, M; Gilbert, JR; Schnetz-Boutaud,

N; Agarwal, A; Postel, EA; Pericak-Vance, MA.Complement factor H variant increases the risk of age-related macular degeneration. *Science,* 2005;308:419-421.

[222] Lommatzsch, A; Hermans, P; Muller, KD; Bornfeld, N; Bird, AC; Pauleikhoff, Are low inflammatory reactions involved in exudative age-related macular degeneration? Morphological and immunhistochemical analysis of AMD associated with basal deposits. *Graefes Arch. Clin. Exp. Ophthalmol.,* 2008;246(6):803-810.

[223] Grossniklaus, HE; Cingle, KA; Yoon, YD; Ketkar, N; L'Hernault, N; Brown, S. Correlation of histologic 2-dimensional reconstruction and confocal scanning laser microscopic imaging of choroidal neovascularization in eyes with age-related maculopathy. *Arch. Ophthalmol.,* 2000;118(5):625-629.

[224] Hinton, DR; He, S; Lopez, PF. Apoptosis in surgically excised choroidal neovascular membranes in age-related macular degeneration. *Arch. Ophthalmol.,* 1998;116(2):203-209.

[225] Killingsworth, MC; Sarks, JP; Sarks, SH. Macrophages related to Bruch's membrane in age-related macular degeneration. *Eye*, 1990;4 (Pt 4):613-621.

[226] Penfold, PL; Killingsworth, MC; Sarks, SH. Senile macular degeneration: the involvement of immunocompetent cells. *Graefes Arch. Clin. Exp. Ophthalmol.*, 1985;223(2):69-76.

[227] Penfold, P; Killingsworth, M; Sarks, S. An ultrastructural study of the role of leucocytes and fibroblasts in the breakdown of Bruch's membrane. *Aust. J. Ophthalmol.*, 1984;12(1):23-31.

[228] Lopez, PF; Grossniklaus, HE; Lambert, HM; Aaberg, TM; Capone, A; Sternberg, P Jr; L'Hernault, N. Pathologic features of surgically excised subretinal neovascular membranes in age-related macular degeneration. *Am. J. Ophthalmol.*, 1991; 112(6):647-656.

[229] Ibrahim, M; Chair, B; Kartz, D. The injured cells: the role of dendritic cell system as a sentinel receptor pathway. *Immunol. Today,* 1995;16:181-186.

[230] Matyszak, M; Perry, V. The potential role of dendritic cells in immune-mediated inflammatory diseases in the central nervous system. *Neuroscience,* 1996;74:599-608.

[231] Uetama, T; Ohno-Matsui, K; Nakahama, K; Morita, I; Mochizuki, M. Phenotypic change regulates monocyte chemoattractant protein-1 (MCP-1) gene expression in human retinal pigment epithelial cells. *J. Cell Physiol.,* 2003;197:77-85.

[232] Higgins, GT; Wang, JH; Dockery, P; Cleary, PE; Redmond, HP. Induction of angiogenic cytokine expression in cultured RPE by ingestion of oxidized photoreceptor outer segments. *Invest. Ophthalmol. Vis. Sci.,* 2003;44:1775-1782.

[233] Ambati, J; Anand, A; Fernandez, S; Sakurai, E; Lynn, BC; Kuziel, WA; Rollins, BJ; Ambati, BK. An animal model of age-related macular degeneration in senescent Ccl-2- or Ccr-2-deficient mice. *Nat. Med.,* 2003; 9: 1390-1397.

[234] Tuo, J; Bojanowski, CM; Zhou, M; Shen, D; Ross, RJ; Rosenberg, KI; Cameron, DJ; Yin, C; Kowalak, JA; Zhuang, Z; Zhang, K; Chan, CC. Murine ccl2/cx3cr1 deficiency results in retinal lesions mimicking human age-related macular degeneration. *Invest. Ophthalmol. Vis. Sci.,* 2007; 48: 3827-3836.

[235] Holtkamp, GM; De Vos, AF; Peek, R; Kijlsta, A. Analysis of the secretion pattern of monocyte chemotactic protein-1 (MCP-1) and transforming growth factor-beta 2 (TGF-beta2) by human retinal pigment epithelial cells. *Clin. Exp. Immunol.,* 1999;118: 35-40.

[236] Zheng, J; Singh, S; Fan, W; Mclaughlin, B. Gene expression in human retinal pigment epithelial cells (ARPE-19) during wound healing. *Exp. Eye Res. (Suppl),* 2000;71:S113.

[237] Pons, M; Marin-Castano, ME. Cigarette smoke-related hydroquinone dysregulates MCP-1, VEGF and PEDF expression in retinal pigment epithelium in vitro and in vivo. *PLoS ONE,* 2011; 28;6(2):e16722.

[238] , S; Samardzija, M; Wenzel, A; Thiersch, M; Grimm, C. Nonessential role of beta3 and beta5 integrin subunits for efficient clearance of cellular debris after light-induced photoreceptor degeneration. *Invest. Ophthalmol. Vis. Sci.,* 2009;50:423-1432.

[239] Ohno-Matsui, K; Morita, I; Tombran-Tink, J; Mrazek, D; Onodera, M. Novel mechanism for age-related macular degeneration: an equilibrium shift between the angiogenesis factors VEGF and PEDF. *J. Cell Physiol.,* 2001. 189:323-333.

[240] Adamis, AP, Shima, DT; Yeo, KT; Yeo, TK; Brown, LF; Berse, B; D'Amore, PA; Folkman, J. Synthesis and secretion of vascular permeability factor/vascular endothelial growth factor by human retinal pigment epithelial cells. *Biochem. Biophys. Res. Commun.,* 1993;193:631-638.

[241] Husain, D; Ambati, B; Adamis, AP; Miller, JW. Mechanisms of age-related macular degeneration. *Ophthalmol. Clin. North Am.,* 2002;15: 87-91.

[242] Kwak, N; Okamoto, N; Wood, JM; Campochiaro, PA. VEGF is major stimulator in model of choroidal neovascularization. *Invest. Ophthalmol. Vis. Sci.*, 2000;41:3158-3164.

[243] Tong, JP; Yao, YF. Contribution of VEGF and PEDF to choroidal angiogenesis: a need for balanced expressions. *Clin. Biochem.*, 2006;39:267-276.

[244] Witmer, AN; Vrensen, GF; Van Noorden, CJ; Schlingemann. RO. Vascular endothelial growth actors and angiogenesis in eye disease. *Prog. Retin. Eye Res.*, 2003;22:1-29.

[245] Kliffen, M; Sharma, HS; Mooy, CM; Kerkvliet, S; de Jong, PT. Increased expression of angiogenic growth factors in age-related maculopathy. *Br. J. Ophthalmol.*, 1997;81:54-162.

[246] Kvanta, A; Algvere, PV; Berglin, L; Seregard, S. Subfoveal fibrovascular membranes in age-related macular degeneration express vascular endothelial growth factor. *Invest. Ophthalmol. Vis. Sci.*, 1996;37:1929-1934.

[247] Lopez, PF; Sippy, BD; Lambert, HM; Thach, AB; Hinton, DR. Transdifferentiated retinal pigment epithelial cells are immunoreactive for vascular endothelial growth factor in surgically excised age-related macular degeneration-related choroidal neovascular membranes. *Invest. Ophthalmol. Vis. Sci.*, 1996;37:855-868.

[248] Reich, SJ; Fosnot, J; Kuroki, A; Tang, W; Yang, X; Maguire, AM; Bennett, J; Tolentino, MJ. Small interfering RNA (siRNA) targeting VEGF effectively inhibits ocular neovascularization in a mouse model. *Mol. Vis.*, 2003;9 210-216.

[249] Olsson, AK; Dimberg, A; Kreuger, J; Claesson-Welsh, L. VEGF receptor signalling - in control of vascular function. *Nat. Rev. Mol. Cell Biol.*, 2006;7: 59-371.

[250] Blaauwgeers, HG; Holtkamp, GM; Rutten, H; Witmer, AN; Koolwijk, P; Partanen, TA; Alitalo, K; Kroon, ME; Kijlstra, A; van Hinsbergh, VW; Schlingemann, RO. Polarized vascular endothelial growth factor secretion by human retinal pigment epithelium and localization of vascular endothelial growth factor receptors on the inner choriocapillaris. Evidence for a trophic paracrine relation. *Am. J. Pathol.*, 1999;155:421-428.

[251] Yi, X; Ogata, N; Komada, M; Yamamoto, C; Takahashi, K; Omori, K; Uyama, M. Vascular endothelial growth factor expression in choroidal neovascularization in rats. *Graefes Arch. Clin. Exp. Ophthalmol.*, 1997;235:313-319.

[252] Dawson, DW; Volpert, OV; Gillis, P; Crawford, SE; Xu, H; Benedict, W; Bouck, NP. Pigment epithelium-derived factor: a potent inhibitor of angiogenesis. *Science*, 1999;285:245-248.

[253] Bhutto, IA; McLeod, DS; Hasegawa, T; Kim, SY; Merges, C; Tong, P; Lutty, GA. Pigment epithelium-derived factor (PEDF) and vascular endothelial growth factor (VEGF) in aged human choroid and eyes with age-related macular degeneration. *Exp. Eye Res.,* 2006;82:99-110.

[254] Ogata, N; Wada, M; Otsuji, T; Jo, N; Tombran-Tink, J; Matsumura, M. Expression of pigment epithelium-derived factor in normal adult rat eye and experimental choroidal neovascularization. *Invest. Ophthalmol. Vis. Sci.,* 2002;43:1168-1175.

[255] Holekamp, NM; Bouck, N; Volpert, O. Pigment epithelium-derived factor is deficient in the vitreous of patients with choroidal neovascularization due to age-related macular degeneration. *Am. J. Ophthalmol.,* 2002;134:220-227.

[256] Apte, RS; Barreiro, RA; Duh, E; Volpert, O; Ferguson, TA. Stimulation of neovascularization by the anti-angiogenic factor PEDF. *Invest. Ophthalmol. Vis. Sci*., 2004;45:4491-4497.

[257] Ohno-Matsui, K; Yoshida, T; Uetama, T; Mochizuki, M; Morita, I.Vascular endothelial growth factor upregulates pigment epithelium-derived factor expression via VEGFR-1 in human retinal pigment epithelial cells. *Biochem. Biophys. Res. Com*mun., 2003;303:962-967.

[258] Fujihara, M; Nagai, N; Sussan, TE; Biswal, S; Handa, JT. Chronic cigarette smoke causes oxidative damage and apoptosis to retinal pigmented epithelial cells in mice *PLoS One*, 2008;3(9):e3119.

In: Hydroquinone
Eds: F. Gokden and A. Lazzarotto
ISBN: 978-1-62100-258-1

Chapter 2

HYDROQUINONE SOLUBILITY AND SEPARATION PROCESSES

Fátima L. Mota[1], António J. Queimada[2] and Eugénia A. Macedo[2]
[1]Université de Toulouse; INPT, UPS;
Laboratoire de Génie Chimique; Toulouse, France
[2]LSRE/LCM - Laboratory of Separation and Reaction Engineering, Faculdade de Engenharia, Universidade do Porto, Porto, Portugal

ABSTRACT

In chemical process design, separation unitsare usually very important for the purification of one or more components from the reaction stage. This step may involve different fluid or solid phases. Chemical engineering thermodynamics is the area of chemical engineering that studies the distribution of such components among equilibrium phases given temperature, pressure and feed composition. Therefore, reliable and accurate data on phase equilibria are crucial to optimize and scale up process separation units. As the number of real conditions in which the interest may be focused in is extremely high, thermodynamic models are frequently used to provide estimates of such data. Among these models,those based on the excess Gibbs energy, regularly known as activity coefficient models, and equations of stateare the most widely used. In this work, an overview of some thermodynamic models to describe the phase equilibria of hydroquinone and its isomers is provided, with the focus on the conditions commonly used for its separation processes. Results on the solid-liquid equilibria in water and organic solvents as a function of temperature, as well as in supercritical carbon dioxide as a function of pressure at several temperatures will be presented using the Non-Random Two Liquid Segment Activity Coefficient model (NRTL-SAC), the UNIFAC, the A-UNIFAC and the Cubic-plus-Association equation of state (CPA).

List of Symbols

A	Helmholtz energy (J)
a	binary interaction parameter (UNIFAC and A-UNIFAC); energy parameter (CPA EoS); constant (Chrastilmodel)
a_0	parameter (CPA EoS)
b	co-volume parameter (CPA EoS); constant (Chrastilmodel)
c_1	parameter (CPA EoS)
g	simplified hard-sphere radial distribution function (CPA EoS)
k	association number (Chrastilmodel)
k_{ij}	binary interaction parameter (CPA EoS)
M	molecular weight (g/mol)
n	total number of moles (mol)
n_I	number of moles of component I (mol)
N_k	number of moles of associating group k (mol)
P	pressure (Pa)
q	constant (Chrastilmodel)
$r_{m,I}$	number of segments of type m present in component I
r_I	total segment number of component I
R	ideal gas constant (8.314 J/mol.K)
S	solubility
T	absolute temperature (K)
X	hydrophobic segment (NRTL-SAC model)
x_s	mole fraction solubility of the solute
X^{A_k}	fraction of non-bonded sites in the solution
$X_I^{A_k}$	fraction of non-bonded sites in pure-component I
$Y-$	polar attractive segment (NRTL-SAC model)
$Y+$	polar repulsive segment (NRTL-SAC model)
Z	hydrophilic segment (NRTL-SAC model)

ABBREVIATIONS

A-UNIFAC	association UNIFAC
AAD	Absolute average deviation (%) $= \frac{1}{NP}\frac{\lvert x^{\exp} - x^{calc}\rvert}{x^{\exp}} \times 100$
CPA	cubic-plus-association
EoS	equation of state
NAG	number of associating groups
NP	number of data points
NRHB	nonrandom hydrogen bonding equation of state
NRTL	nonrandom two liquid
SAC	segment activity coefficient
SAFT	statistical associating fluid theory
VLE	vapor-liquid equilibrium
UNIFAC	UNIversal quasi chemical Functional group Activity Coefficient method
vdWV	van der Waals volume (m^3/mol)

Greek Symbols

β_{ij}	cross-association volume
$\beta^{A_iB_j}$	association volume (CPA EoS)
γ	activity coefficient
Γ_m^{lc}	activity coefficient of segment species *m*(NRTL-SAC model)
$\Gamma_m^{lc,I}$	activity coefficient of segment species *m* contained only in component *I* (NRTL-SAC model)
δ	parameter (CPA EoS)
$\Delta^{A_kB_j}$	associating strength between site *A* of group *k* and site *B* of group *j* (A-UNIFAC method and CPA EoS)
ΔC_p	difference of the liquid and solid molar heat capacities (J/mol.K)
ΔH	heat of dissolution (J/mol)
$\Delta_{fus}H$	enthalpy of fusion (J/mol)
$\Delta_{sol}G$	molar Gibbs energy of dissolution (J/mol)
$\Delta_{sol}H$	molar enthalpy of dissolution (J/mol)

$\Delta_{sol}S$	molar entropy of dissolution (J/mol.K)
$\varepsilon^{A_kB_j}$	association energy (J/mol) (A-UNIFAC method and CPA EoS)
η	reduced fluid density (CPA EoS)
$\kappa^{A_kB_j}$	association volume
ϑ_k^I	number of groups of type *k* in a molecule of component *I*
ρ	density
ρ_j	associating group density in the mixture (A-UNIFAC method)
$(\rho_j)_I$	density of group *j* in pure component *I* (A-UNIFAC method)
φ	fugacity coefficient
ϕ	segment mole fraction (NRTL-SAC model)

Subscripts

I, J	component indices
i, m	segment based species indexes(NRTL-SAC model); component index (CPA EoS)
c	critical; solvent
m	melting
r	reduced
S	solute
0	pure component

Superscripts

Assoc	association
A_k	associating site of group *k*
C	combinatorial
Calc	calculated
exp	experimental
lc	local composition
R	residual

1. Introduction

Phenolic compounds are a chemical family whose members have one or more hydroxyl groups attached to an aromatic ring. The interest on these molecules has grown greatly, with special attention focused on those with food or medical applications. Hydroquinone is one of those cases, in spite of recent studies associating this molecule to carcinogenic effects [1].

Hydroquinone is an important organic compound and has been extensively used in the fields of pharmaceuticals, food, and industrial chemicals [2]. Its most important applications are related with its reducing power. It is frequently used as a reducing agent, as a photographic developer, as an antioxidant for many unstable oxidizable products, as a stabilizer or polymerizing inhibitor for certain materials that polymerize in the presence of free radicals, and as a chemical intermediate for the production of antioxidants, antiozonants, agro-chemicals, and polymers. It is a skin-lightening agent and is used as well in cosmetics and medical preparations [3]. Hydroquinone is the most produced benzenediol with an annual world production of 40000 – 50000 ton (estimated for the years between 1990 and 2000). Nearly 90% of the produced hydroquinone in the USA is used domestically: 40% as photographic developer for film and paper, 40% as antioxidant and 15% as polymerization inhibitor.

Among the four principal routes to prepare dihydroxybenzenes, only three are selective for hydroquinone: oxidation of aniline, oxidation of p-diisopropylbenzene, and hydroxylation of phenol by hydrogen peroxide (simultaneous synthesis of hydroquinone and catechol, respectively the *para* and *ortho* isomers of dihydroxybenzene).

Usually, cooling crystallization is used to purify hydroquinone from the reactive mixtures. The crystal habit is an important factor to be considered during the crystallization process, because it greatly affects the product application and following processes (filtering, washing, drying, packaging, storage and transportation). However, the crystal habit is evidently influenced by the solvent used in the crystallization. Therefore, to select the appropriate solvents, an adequate knowledge of solubility is required.

A mixture of resorcinol (1,3-dihydroxybenzene) and hydroquinone is sometimes found at the output of hydroquinone production processes. These compounds have very close boiling points (559 Kand 550 K, respectively) and high sublimation points, which makes distillation a difficult separation process. Separation by crystallization is thus a more feasible alternative, provided the mixture has an eutectic point, being possible to isolate one of the compounds in a pure form. In fact, it was found that this mixture could be efficiently separated by adding a particular organic solvent [4,5].

In the chemical industry, most of the separation and reaction processes require the knowledge of the phase equilibria of the involved molecules in pure, mixed orsupercritical solvents. In this way, the fate and separation processes can be better understood. Besides the importance of systems at ambient pressure conditions, many recent works regarding the estimation of the phase equilibriaof biological molecules are concerned with their supercritical fluid processing. This reveals to be very important because in the food and pharmaceutical industries, many molecules cannot be processed at high temperatures due to their decomposition. Besides the inherent complexities related with experimental measurements, theachievement of accurate and reliable results is especially difficult. Therefore, theoretical approaches to predict phase equilibria are essential.

In this sense, several thermodynamic models were applied to describe the phase equilibria of hydroquinone in water, alcohols, esters and carboxylic acids. The solubility data in supercritical carbon dioxide is also addressed. Excess Gibbs energy models, like the group-contribution UNIFAC and A-UNIFAC (UNIFAC with an explicit association term), and the NonRandom Two-Liquid Segment Activity Coefficient (NRTL-SAC) models were compared to an equation of state (EoS), the Cubic-plus-Association (CPA). The predictive group-contribution methods, whichbuild molecules from their constituent functional groups, require only the chemical structure information for the solute and solvents, and the group-interaction parameters. The NRTL-SAC model represents molecules by conceptual segments, and requires only a set of well-chosen data to fit a small number of parameters. CPA combines the Soave-Redlich-KwongEoSto describe the physical interactions with the Wertheim's first order perturbation theory, which can be applied to different types of hydrogen-bonding compounds. As hydroquinone is associative and, in general, the solvents studied are also associative, models specially dealing with these interactions are more appropriate.

To apply thermodynamic models, critical and melting properties were collected from the literature. For the UNIFAC model, no parameter estimation wasrequired. For A-UNIFAC, a reparameterization was necessary to obtain the group-interaction parameters for the –ACOH group (aromatic hydroxyl) interacting with water, alcohol, ester and acid groups. For NRTL-SAC, the segment numbers (polar, hydrophobic and hydrophilic) were calculated based on solubility data in six solvents at 298.15 K. For CPA, the cubic parameters and the energy and volume of association were obtained from regression of vapor pressure and liquid density data.

Very good results were achieved for the solid-liquid equilibrium of hydroquinone in these solvents using any of the above mentioned models, being the CPA EoS and the NRTL-SAC, the best. Regarding the group-contribution methods, UNIFAC performed slightly better, except in water, where A-UNIFAC wassuperior, showing the importance of taking association effects into account.

2. Manufacturing and Processing

Phenol is the structure upon which the entire group of phenolicsis based. In many ways, they are similar to alcohols of aliphatic structures where the hydroxyl group is, however, influenced by the presence of the aromatic ring. Because of that, the hydrogen of the phenolic hydroxyl is labile, which makes phenols weak acids. Phenolic compounds are characteristic of plants and, as a group, they are usually found as esters or glycosides rather than as free compounds. Hydroquinone (1,4-benzenediol), or quinol, is an aromatic organic compound in which the two hydroxyl groups are bonded to the aromatic ring in a *para* position. There are two other isomers of benzenediol: 1,2-benzenediol (*ortho*), commonly known as catechol, and 1,3-benzenediol (*meta*), commonly known as resorcinol.

Hydroquinone occurs in the environment as a result of anthropogenic sources, as well as in natural products from plants and animals.However, bioaccumulation has not been observed, as it readily biodegrades and photodegrades [6,7]. Hydroquinone has been found in plant-derived food products e.g., wheat germ, in brewed coffee, and in teas prepared from the leaves of some berries, where the concentration sometimes exceeds 1% [3]. Catechol occurs naturally in kino and in beechwood tar and, in small amounts, in fruits and vegetables [8]. Resorcinol has been found in a wide variety of natural products: broad bean, honey mushroom, and exudates of seedlings of the yellow pond lily, and additionally in extracts of tobacco leaves, being thus found as a component of tobacco smoke [9].The benzenediols are industrially prepared according to four different routes: (1) oxidation of aniline, (2) alkali fusion of m-benzenedisulfonic acid, (3) oxidation of p- or m-diisopropylbenzene, and (4) hydroxylation of phenol by hydrogen peroxidase [10].Route (1) is specificfor hydroquinone, while route (2) is specific for resorcinol. Route (3) is able to produce hydroquinone and resorcinol, while route (4) produces hydroquinone and catechol.In the following,a brief review on the chemistry of these routes is given, to better understand the involved separation processes.

2.1. Hydroperoxidation of P-Diisopropylbenzene

The most widely used route to produce hydroquinoneconsists on the dialkylation of benzene with propene to give 1,4-diisopropylbenzene. This compound reacts with air to produce the bis(hydroperoxidase) and rearranges in acidic media to give acetone and hydroquinone (Figure 1). The reaction is similar for hydroquinone and resorcinol, and the product depends on the diisopropylbenzene isomer that is formed in the first step.

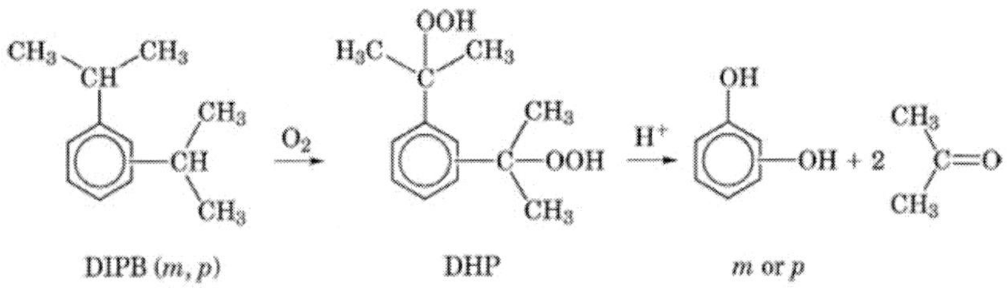

Figure 1. Hydroperoxidation of m- or p-diisopropylbenzene.

2.2. Hydroxylation of Phenol

A second route involves the hydroxylation of phenol by hydrogen peroxideproviding a mixture of hydroquinone and catechol.

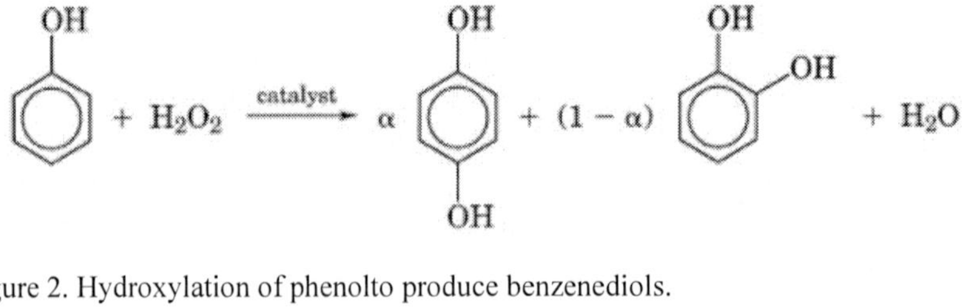

Figure 2. Hydroxylation of phenolto produce benzenediols.

This process has been widely studied and led to the construction of new industrial units, because it is simple and does not have undesirable by-products. However, in order to be economically reliable, high yields of dihydroxybenzenes and a high flexibility for the isomeric ratio hydroquinone to catechol are required.Using homogeneous catalysis; the yields in dihydroxybenzenes based on hydrogen peroxide are generally lower than 60%. The discovery of hydrogen peroxide hydroxylation of phenol in presence of strong acids (perchloric, trifluoromethanesulfonic, or sulfuric) allowed higher yields (85% based on hydrogen peroxide). This reaction can be run without solvents and does not generate resorcinol. Its main advantage relies on the fact that the ratio hydroquinone/ catechol can be tuned. This ratio increases with the acidity of the catalyst used.Using heterogeneous catalysis for the phenol hydroxylation by hydrogen peroxide, yields as high as 88% (based on hydrogen peroxide) or 96% (based on phenol) can be obtained, but at the expense of more restricted conditions (for example, titanium silicate at 80 °C in 2h) and solvents (for example, acetone).

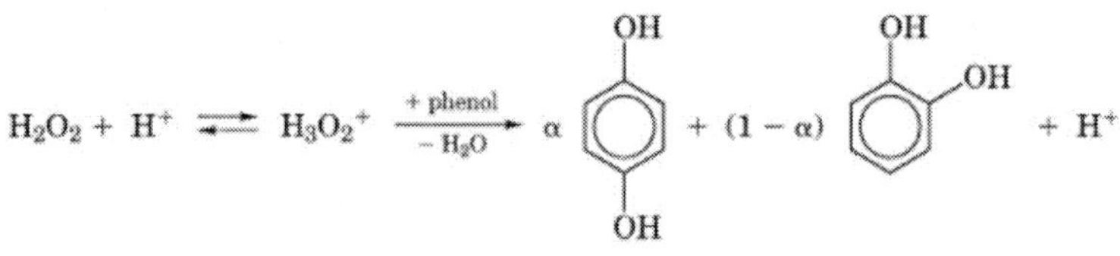

Figure 3. Hydroxylation of phenol in the presence of strong acids.

2.3. Aniline Oxidation

The third method, and the only exclusively selective for hydroquinone, is the oxidation of aniline by manganese dioxide in aqueous sulfuric acid. The resulting benzoquinone, isolated by vapor stripping, is reduced in a second step by either an aqueous acidic suspension of iron metal or by catalytic hydrogenation.

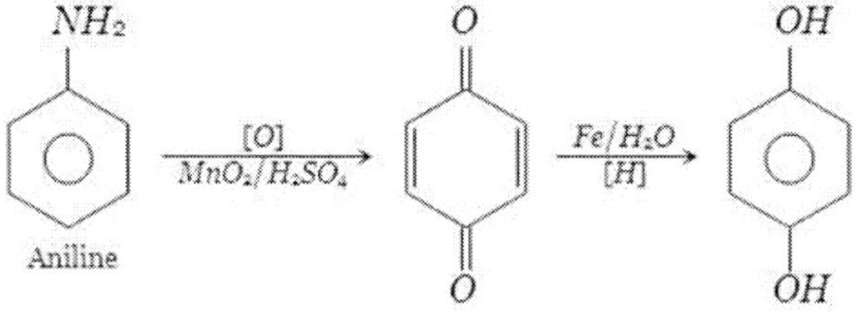

Figure 4. Oxidation of aniline.

The yield is 85 to 90% of hydroquinone. The process is conducted batchwise and generates substantial waste products: manganese sulfate, ammonium sulfate and iron oxides which are environmentally harmful. This is a quite old process having limited commercial use and did not show any improvement since 1980. Even though, manganese sulfate is being used in the agricultural field to promote plant growth, and ammonium sulfate as a spray adjuvant for water soluble insecticides, herbicides and fungicides.Few other solutions have been developed to dispose these unsuitable by-products.

2.4. Other Routes

Routes starting from o- or p-dichlorobenzene are no longer used due to the formation of sodium chloride during the sodium hydroxide hydrolysis.A method starting from bisphenol A allows a selective access to hydroquinone:

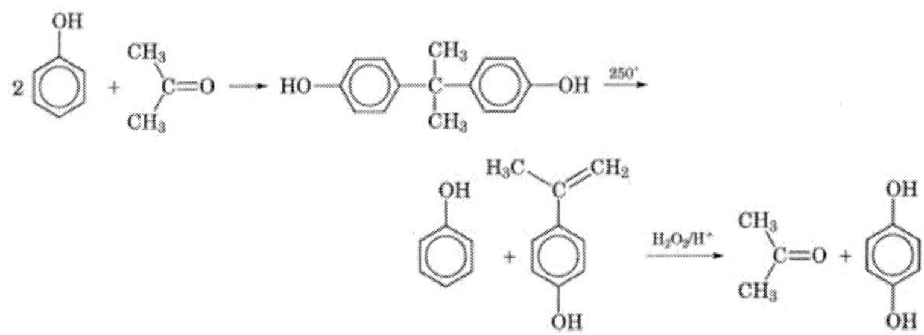

Figure 5. Production of hydroquinone from bisphenol A.

In spite of the obtained yield of hydroquinone being close to 90%, and the phenol and acetone formed being recoverable for a following reactive process,this process was not yet industrialized.A variety of other methods can be found, but not so relevant as the ones presented above.

3. Purification and Solubility

Process modeling is an important tool to evaluate in advance if a given separation will be technically feasible or more economically sound. To do that, a process engineer uses a set of modelsto check if the selected process configuration is the most adequate. During this preliminary stage,the process engineer will work on the process optimization to minimize costs and maximize products, while fulfilling the quality requirements.

A critical point in process modeling is related with the right description of intermolecular interactions. This field is usually dealt in within chemical engineering by applied thermodynamics. In this research field, experimental measurements and computational modeling are combined so that the best description of the relevant systems is achieved. For example, having an adequatethermodynamic model, a given feed composition, temperature and pressure, the number of equilibrium phases, phase fractions and compositions can be determined. This information is most useful for some operations such as distillation, liquid-liquid extraction or crystallization processes.

Distillation is a physical separation process based on differences of the volatilities in a boiling liquid mixture. It has been widely used as the separation process of choice in the chemical industry because of its reliability, simplicity and low capital cost. However, distillation systems have also relatively high energy consumption. If the difference in the vapor pressure between two components A and B is large (generally expressed as the difference in boiling points), the mixture in the beginning of the distillation is highly enriched in component A, and when component A has been distilled off, the boiling liquid is enriched in component B. Resorcinol and hydroquinone have close boiling points (Table 1) and, in addition, they both have a high sublimation enthalpy, and therefore they cannot be easily separated from each otherby distillation, in a pure form.

Most chemicals, such as hydroquinone, are solids at ambient conditions. Thus, distillation processes are mostly used to evaporate solvents, while liquid-liquid extraction is regularly used for separation processes.

Crystallization is indeed the most important separation and purification process employed to produce a broad range of materials, such as in the fine chemicals, food and pharmaceutical industries. It influences the performance of downstream operations such as filtering, washing, drying, package, storage and transportation, and the physical-chemical properties of the final product depend on the crystal properties. Usually, cooling crystallization is used to purify such compounds, but crystallization by addition of solvents that reduce the solubility (antisolvent crystallization) is also common. The crystal habit is an important factor needing to be considered in the crystallization process because it greatly affects the product application and following processes [11]. As for the majority of organic compounds, the solubility and crystal habit are evidently influenced by the solvent used in the crystallization, an adequate knowledge of phase equilibria is required to select the appropriate solvents.

Important properties interfering in the solubilization of solids in liquids are the fusion enthalpies ($\Delta_{fus}H$) and melting temperatures (T_m), as well as the thermal heat capacity (ΔC_p), as will be latter explained based on Eq. 4. These properties for the three isomers of dihydroxybenzene are presented in Table 1, as well as for phenol for comparison purposes.Hydroquinone is the one requiring the highest temperature, and energy, to break the crystal structure, and so, most of the times, it is also the least soluble.

Table 1. Physical properties of mono- and dihydroxy- benzenes [12]

	Hydroquinone	Resorcinol	Catechol	Phenol
T_m (^{0}C)	171.6	108.9	104.5	40.91
$\Delta_{fus}H$ (kJ/mol)	27.11	21.00	22.76	11.51
ΔC_p(J/mol.K)	55.19	60.59	10.85	99.21
T_b (^{0}C)	286.0	276.5	230.7	181.84

To gain an insight into the phase behavior of the related mixtures and to evaluate the possibility of solvent recovery, a series of vapor-liquid equilibrium (VLE), solubilities in supercritical carbon dioxide, and solubilities in pure solvents are presented in the next sections.

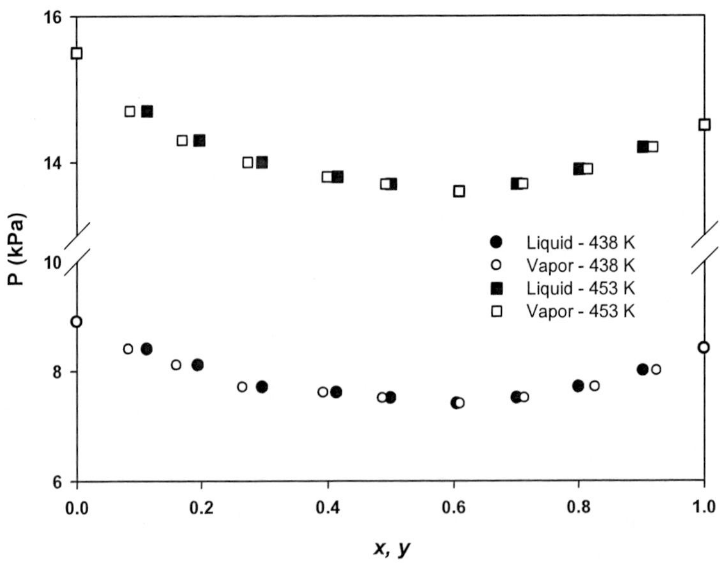

Figure 6. VLE data for 4-methoxyphenol+catechol at 438 and 453 K [13].

3.1. Vapor-Liquid Equilibrium

Experimental vapor-liquid equilibrium (VLE) data concerning hydroquinone and its isomers were found in the literature and are discussed below. The mixture 4-methoxyphenol + catechol have two closely boiling constituents (516 and 519 K, respectively), and may be encountered in the manufacturing of phenol derivatives. Isothermal VLE data are presented for the system4-methoxyphenol+catecholin Figure 6 The binary system exhibits a minimum pressure azeotrope, leading to economic and technical infeasibility for the separation of this mixture by means of conventional distillation [13].

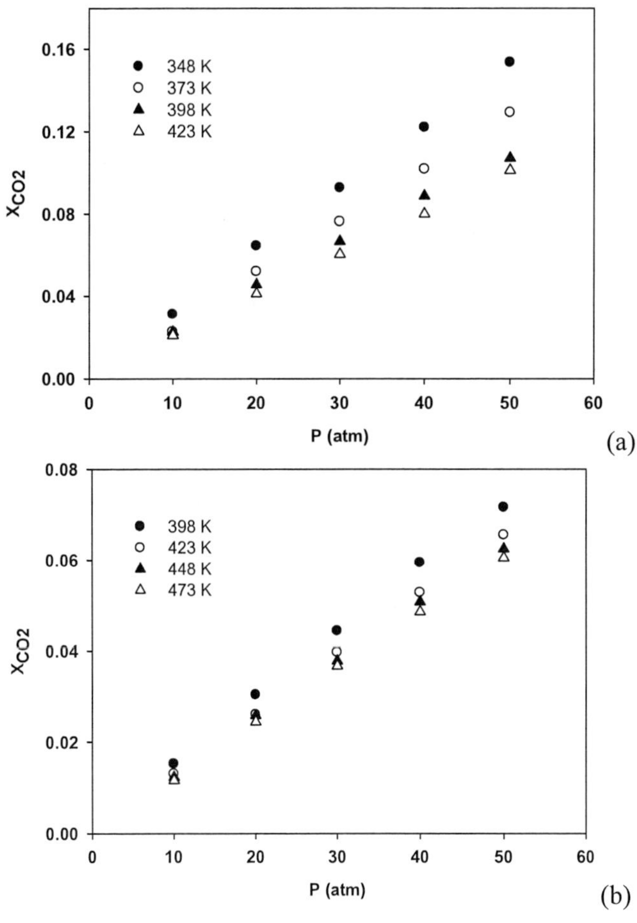

Figure 7. VLE data for carbon dioxide + (a) phenol and (b) catechol systems [14].

Figure 7 presents VLE data for carbon dioxide (gas solubilities) with phenol at 348-423 K, and for catechol at 398-473 K[14]. The solubility of carbon dioxide decreases with rising temperature and increases with pressure, and the carbon dioxide solubilitiesare higher in phenol than in catechol. Thistype of separation processes based on gas solubilities can be easily carried out at industrial scale.

3.2. Supercritical Equilibrium

The main advantages of supercritical fluid technology include increased speed, easy solvent separation and better recovery, and reduction in both solvent usage and solvent waste generation. The most widely used supercritical solvent is carbon dioxide since it is nontoxic, nonflammable and relatively inexpensive. Also, its critical properties are moderate, as well as it has a high solvent power for a wide range of organic compounds. The solubility in supercritical fluids is one of the most important thermophysical properties in supercritical technology, which must be measured or modeled, for an efficient process design. In this way, a set of collected dihydroxybenzene solubility data in supercritical carbon dioxide is discussed below. The solubility of hydroquinone in supercritical carbon dioxide was found to be very low, in the order of 10^{-6} (mole fraction solubility) at 308 K and pressuresranging from 120 to 200 atm [15]. The solubilities of resorcinol and catechol are additionally presented in Figure 8. At a constant temperature, the solubilities of the solutes increase with pressure. However, the temperature plays a much more important role in the solubility than the pressure. The results obtained by these authors indicate that the solubility of the dihydroxybenzene isomers vary in the following order: catechol, resorcinol and the very least soluble hydroquinone [15].In hydroquinone, the intermolecular hydrogen bonding is stronger and so, the solubility is lower.

Solubility measurements of pure hydroquinone in supercritical CO_2 at two temperatures were reported by Coutsikos et al.[16]at pressures rangingfrom 85 to 312 bar, as presented in Figure 9. These solubilities are lower than those of phenol by about three orders of magnitude. This difference is mainly due to the much lower vapor pressure of hydroquinone (0.0102 Pa at 308 K), caused by the two hydroxyl groups, as compared to that of phenol (124.46 Pa at 309 K). Most of the investigations on solubility have been concerned with binary systems consisting of a single solute in contact with a supercritical fluid.

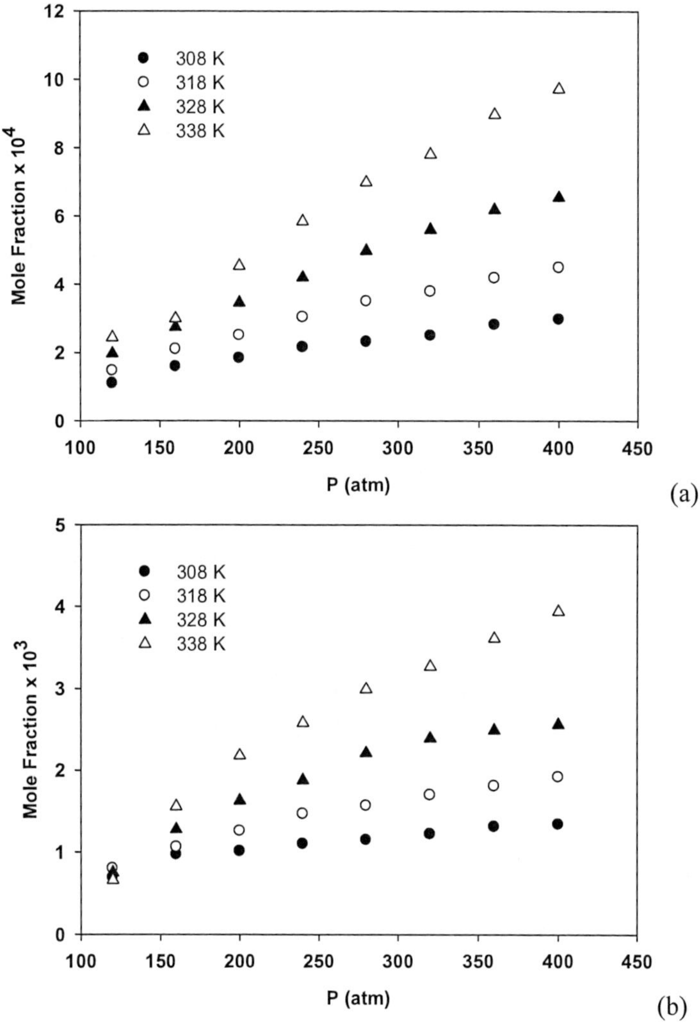

Figure 8. Solubilities of (a) resorcinol and (b) catechol in supercritical carbon dioxide [15].

However, the interactions between solutes can significantly alter solubility in the supercritical fluid phase. The solubilities of hydroquinone and p-quinone and their equimolar mixture in supercritical carbon dioxide over the pressure range from 98.69 to 345.42atm, at 333, 348 and 363 K, were studied by García-González et al.[17], as presented in Figures 10-11. For both solutes, the solubility increased with pressure, at constant temperature.

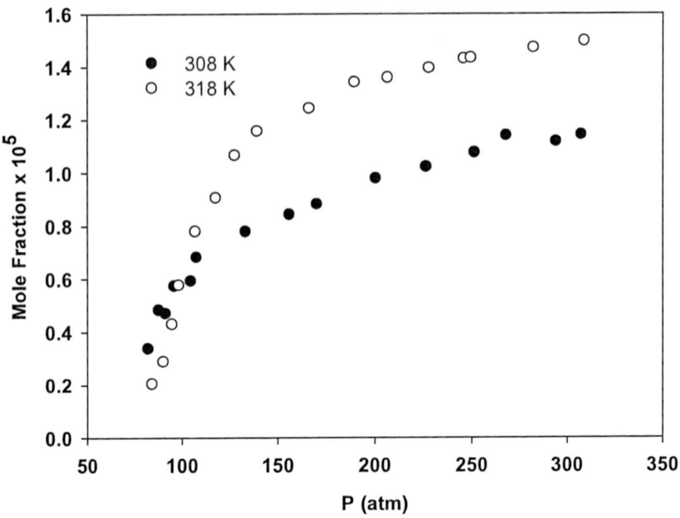

Figure 9. Solubilities of hydroquinonein supercritical carbon dioxide [16].

However, the temperature effect is more complex. For both solutes in binary systems with CO_2, and for hydroquinone in the ternary, there is a pressure range in which the solubility decreases with temperature, while beyond this pressure range the solubility increases.

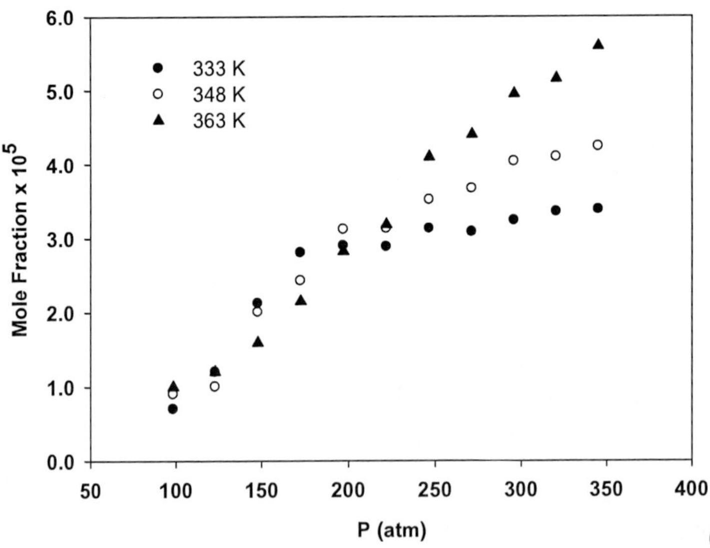

(a)

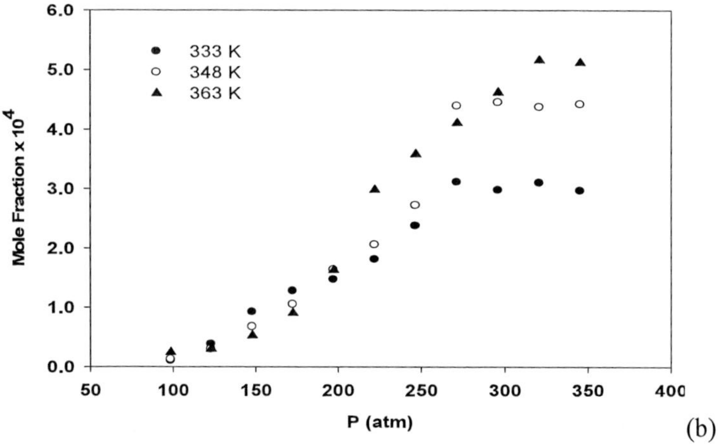

Figure 10. Solubilities of hydroquinone in supercritical carbon dioxide: (a) binary system and (b) ternary system with p-quinoneequimolar mixture [17].

For *p*-quinone in the ternary system, the solubility decreases with temperature in the whole pressure range. *p*-quinoneshows a solubility approximately three orders of magnitude higher than hydroquinone in binary systems. Hydroquinone has strong intermolecular hydrogen bonding and thus, a much higher melting point and lower vapor pressure, which consequently reduces its solubility in the gas phase.

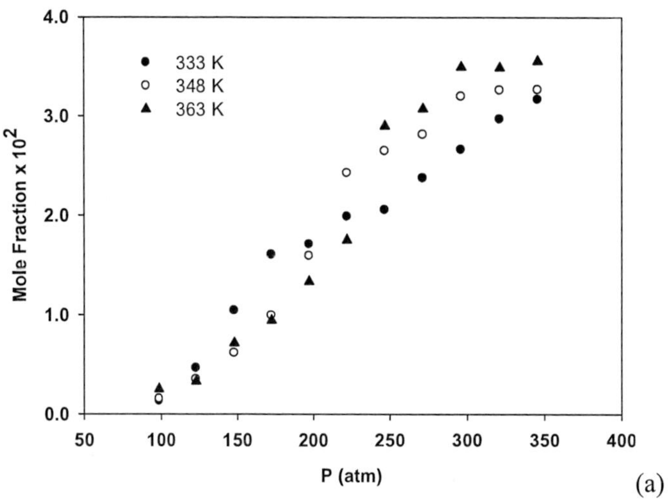

Figure 11. (Continued).

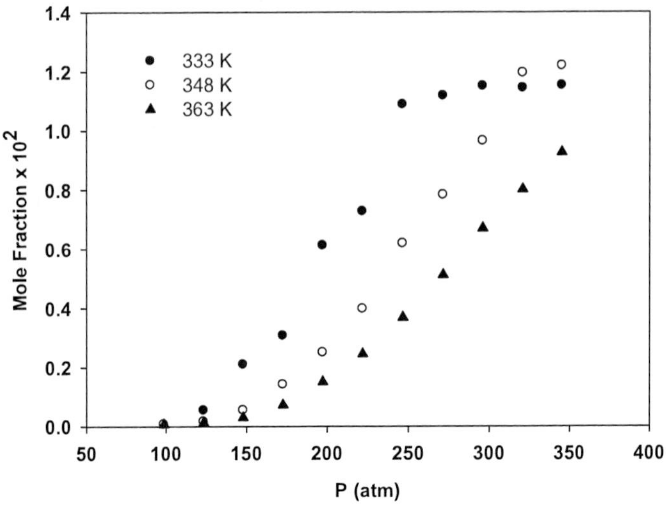

(b)

Figure 11. Solubilities of p-quinone in supercritical carbon dioxide: (a) binary system and (b) ternary system with hydroquinoneequimolar mixture [17].

In the ternary system, *p*-quinone enhances tremendously the solubility of hydroquinone, while its solubility is reduced.

3.3. Solid-Liquid Equilibrium

One of the most widely used methods in purification is crystallization, especially because these compounds are generally solid under normal conditions.So, solid-liquid equilibrium (SLE) is viewed as a key property for crystallization design. In spite of the available solubility data being scarce, some data were collected from the literature, and a discussion on that is presented below.

A solid mixture containing resorcinol and hydroquinone is separated efficiently by mixing with a particular organic solvent such as esters, ethers, alcohols, or vinyl group containing compounds, and separating the resulting slurry mixture into precipitated hydroquinone and dissolved resorcinol[5]. Therefore, theright choice of solvent allows dissolving resorcinol but not hydroquinone. Pure resorcinol and hydroquinone can also be obtained by mixing their solid mixture with 15-30% by weight of methanol (containing 5-50% by weight of water): the final products are obtained by successive fractional crystallization and separation [4].

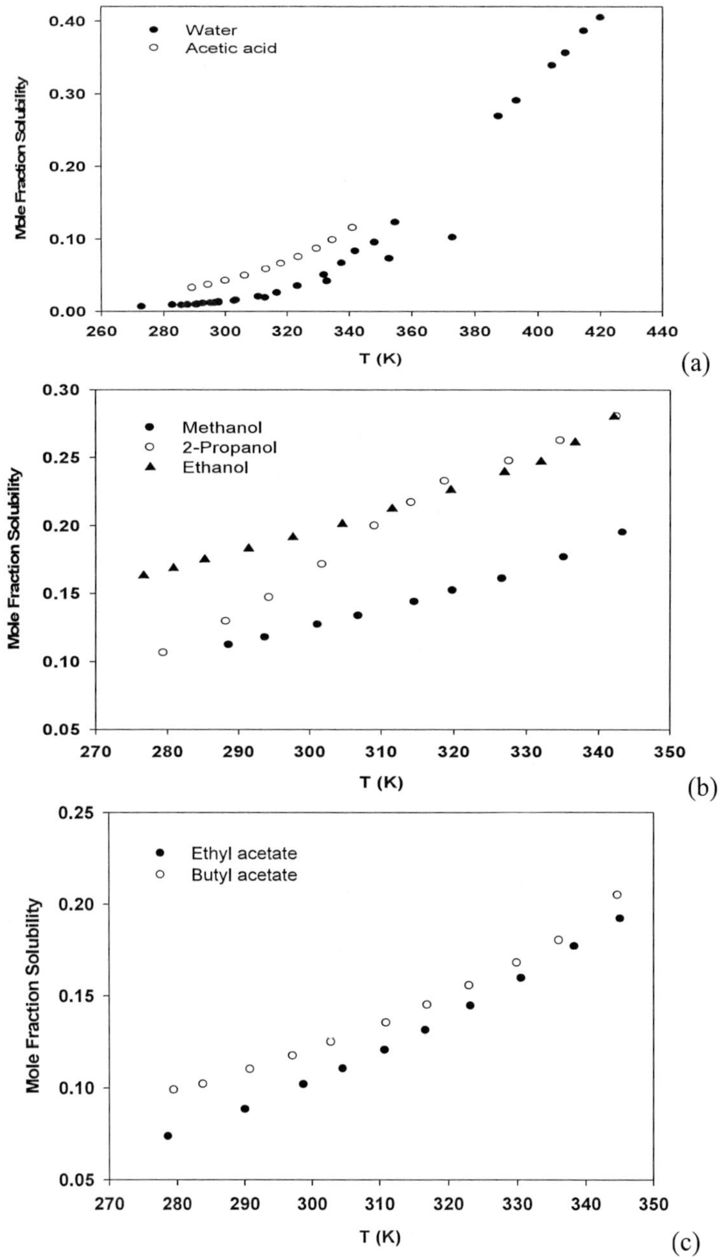

Figure 12. Solubility of hydroquinone in different solvents as a function of temperature: (a) water and acetic acid; (b) alcohols; and (c) esters [2].

Usually, a cooling crystallization process is used to purify hydroquinone. The crystal habit is an important factor needing to be considered because it greatly affects the product applications and further processing (filtering, washing, drying, packing, stirringand transportation).

This crystal habit is influenced by several factors, such as: the solvents used in the crystallization process, presence of impurities or crystalline seeds in solution, solution stirring and cooling rate. The first task in a crystallization process is to choose the appropriate solvents to obtain the desired crystal habit [11].

The solubility is related to the crystal habit: to design the adequate crystallization process to produce the crystals with the adequate habit, solubility values in different solvents are fundamental; in this way, the metastable zone width, the ideal zone to grow crystals, should be known.

Li et al. [2] measured the solubility of hydroquinone in water, methanol, ethanol, 2-propanol, ethyl acetate, butyl acetate and acetic acid, from 277 to 345 K, under atmospheric pressure. The results are presented in Figure 12.

The solubility is higher in alcohols, followed by the esters, acetic acid and the lowestvalues are found in water. In 2-propanol, the solubility curve has a unique behavior perhaps due to complex solute-solvent interactions. For temperatures higher than 360 K, the only data available are the aqueous solubilities [18]. The solubility in all solvents follows the expected approximately exponential increasing trend with temperature.

The aqueous solubilities of catechol and resorcinol can also be found in the literature [18], and are presented in Figure 13.

It is important to note the scattering which, in general, is related to the difficulties inherent to the experimental measurements of small solubilities. Even so, the aqueous solubilities of hydroquinone are in general the lowest, while among 273 and 333 K those of resorcinol are the highest, and for temperatures higher than 333 K, the solubilities of catechol are the highest. Given that the structure of these molecules is the same, being the only difference the position of one hydroxyl group in the aromatic ring, this order follows the extent of hydrogen bonding with water.

The most common way to evaluate the effect of temperature on solubility is the calculation of solubility products at different temperatures. Alternatively, the thermodynamic properties of dissolution can be calculated using experimental data. These parameters reflect the modification of the solution properties due to the presence of the solute at infinite dilution, at a given temperature.

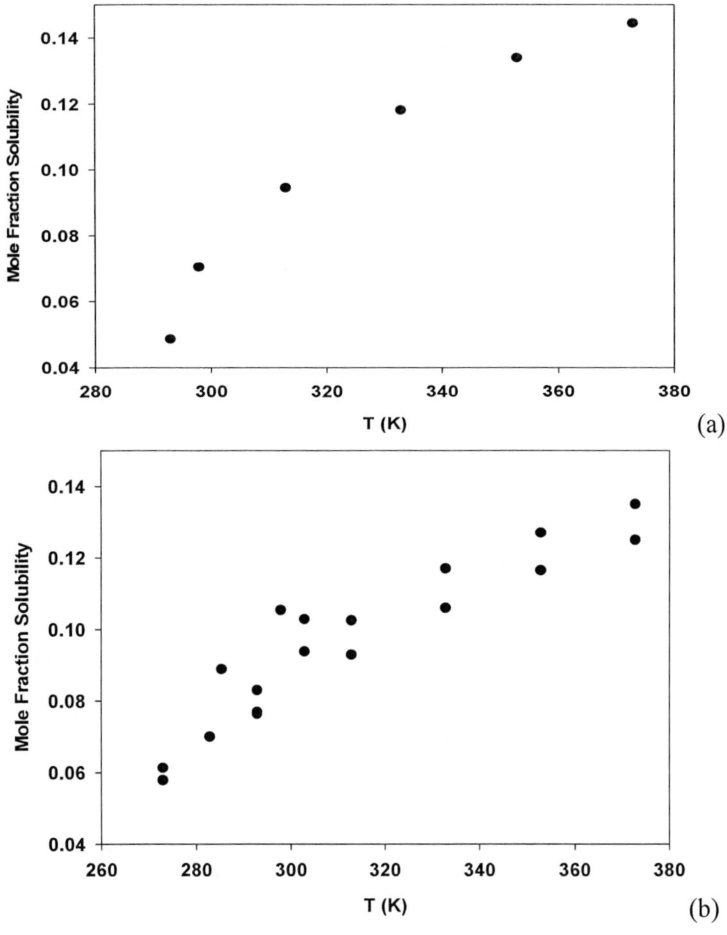

Figure 13. Aqueous solubility as a function of temperature of (a) catechol and (b) resorcinol [18].

Assuming that the activity coefficient of the solute is equal to one in the hypothetical dilute ideal solution, the following equation can be obtained [19]:

$$\Delta_{sol} H = RT^2 \left(\frac{d \ln x_s}{dT} \right)_P \tag{1}$$

where the molar enthalpy of dissolution $\Delta_{sol}H$ is the difference between the partial molar enthalpy of the compound in the solution and the pure molar enthalpy at absolute temperature T, x_s is the mole fraction solubility of the solute, R is the ideal gas constant, and P is the pressure. Representing $ln\ x$ against temperature, $\Delta_{sol}H$ can be obtained. The remaining thermodynamic functions, namely the molar Gibbs energy ($\Delta_{sol}G$) and entropy ($\Delta_{sol}S$), are obtained from the following definitions:

$$\Delta_{sol}G = -RT\ln(x_s)_P \tag{2}$$

$$\Delta_{sol}S = \frac{\Delta_{sol}H - \Delta_{sol}G}{T} \tag{3}$$

Linear solubility-temperature plots are used because they can be interpolated and extrapolated quite accurately, and in an easy way can be treated by the statistics of linear regression, providing thermodynamic data for the dissolution process.

The thermodynamic properties of dissolution of hydroquinone calculated using experimental solubility data are presented in Table 2. For catechol and resorcinol, in spite of data scattering, the following values were found for enthalpy, Gibbs energy and entropy of dissolution, respectively, at 298 K: for catechol, 10.41 kJ/mol, 6.58 kJ/mol and 12.86 J/mol.K; for resorcinol, 5.54 kJ/mol, 6.02 kJ/mol and -1.62 J/mol.K. In spite of having a negative entropic contribution, resorcinol is the most favourably dissolved in water due to a favourable enthalpic contribution.

To better understand the relation between solubility and these properties, in Figure 14 a representation of the enthalpic*versus* entropic contributionsis presented for hydroquinone in organic solvents.

Table 2. Thermodynamic properties of dissolution (Eqs. 1-3) of hydroquinone in several solvents 298 K

Solvent	$\Delta_{sol}H$ (kJ/mol)	$\Delta_{sol}G$ (kJ/mol)	$\Delta_{sol}S$ (J/mol.K)
water	22.96	11.06	39.93
methanol	7.46	5.17	7.63
ethanol	5.82	4.09	5.80
2-propanol	11.73	4.43	24.21
ethyl acetate	10.76	5.68	17.02
butyl acetate	8.39	5.24	10.39
acetic acid	18.50	7.91	35.27

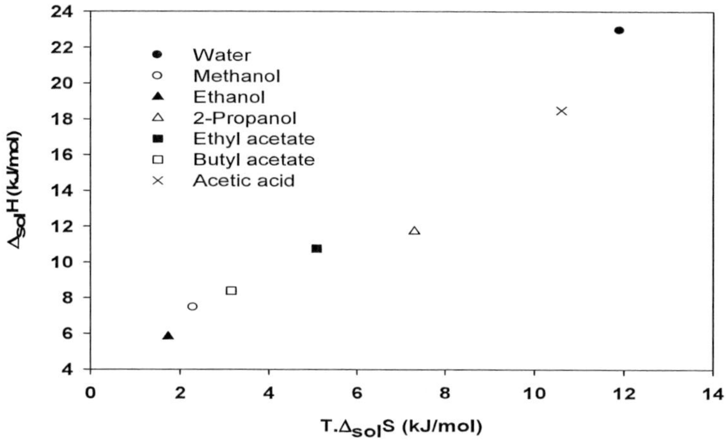

Figure 14. Enthalpy against entropy of dissolution at 298 K for hydroquinone in several solvents.

In terms of enthalpy, hydroquinone is more favourably solubilized in ethanol. In water, it has a strong entropic penalty, and so lowersolubilities. In 2-propanol and ethyl acetate, it has similar enthalpic contributions, but it is more soluble in 2-propanol because of amore favourable entropic contribution.

4. Thermodynamic Modelling

Considering a pure solid phase with no solid-solid phase transitions, assuming the difference in the heat capacities between the liquid and solid phases of the solute (ΔC_p) to be constant in the temperature range [T, T_m], and neglecting the effect of pressure (melting temperature and enthalpy, heat capacities and Poynting term), the solubility of a solute s can be calculated from the following generalized expression that relates the reference-state fugacities [20]:

$$\ln\frac{f_s^{liq}(T,P)}{f_s^{sol}(T,P)}=\frac{\Delta_{fus}H}{R}\left(\frac{1}{T}-\frac{1}{T_m}\right)-\frac{\Delta C_p}{R}\left[\frac{T_m}{T}-\ln\left(\frac{T_m}{T}\right)-1\right] \qquad (4)$$

where $\Delta_{fus}H$ is the solute enthalpy of fusion, and T_m is the solute melting temperature, . Using an activity coefficient model, the following expression for the mole fraction solubility (x_s) is obtained:

$$x_s = \frac{1}{\gamma_s}\exp\left\{-\frac{\Delta_{fus}H}{R}\left(\frac{1}{T}-\frac{1}{T_m}\right)+\frac{\Delta C_p}{R}\left[\frac{T_m}{T}-\ln\left(\frac{T_m}{T}\right)-1\right]\right\} \qquad (5)$$

where γ_s is the solute activity coefficient. When an equation of state is selected:

$$x_s = \frac{\varphi_s^{liq_0}}{\varphi_s^{liq}}\exp\left\{-\frac{\Delta_{fus}H_s}{R}\left(\frac{1}{T}-\frac{1}{T_m}\right)+\frac{\Delta C_p}{R}\left[\frac{T_m}{T}-\ln\left(\frac{T_m}{T}\right)-1\right]\right\} \qquad (6)$$

where φ is the fugacity coefficient and the subscript *0*refers to a pure component.

If the fugacity ratio or the activity coefficient is equal to unity, the solution is considered ideal, but the majority of the solutions are far from ideality. Different models can be used to calculate these variables accounting for non-idealities.

Some empirical correlations based on physical-chemical properties, like the octanol-water partition coefficient and the melting point, were proposed to estimate solubilitiesbut they are currently of little use [21,22]. To calculate the partition coefficient several reliable methods are known, but that is not the case for the melting point. The so-called log-linear model [23]which uses the partition coefficient cannot predict solubility in systems that present maxima, and this is a reason why these empirical models should be replaced by more theoretically sound models, such as the thermodynamic models regularly used for phase equilibria.

The Scatchard-Hildebrand concept of regular solutions [20] has been one of the thermodynamic models most used in the pharmaceutical and chemical industries, and there are several modifications. According to this theory, the existence of a maximum in the solubility curve is observed when the solvent and solute solubility parameters are the same. A drawback of this model is its very simple assumptions which lead to very limited practical use in the prediction of the solubility of complex substances. Some other activity coefficient models have also been proposed, but not all are practical because phase equilibrium data are frequently not available and the application of these models regularly requires the determination of binary interaction parameters from experimental data.

Classical equations of state (EoS) are seldomly used to obtain the solubility of drug-like molecules, either because these are typically highly non-ideal systems, or the solute complexity is so high that it is often impractical to determine the model parameters from the reduced data sets available. The limitations of the excess Gibbs energy models for high pressure and temperature phase equilibria, non-ideal vapour phases as well as their need of mixture data, suggests further research on the application of EoS for these systems. The statistical associating fluid theory (SAFT) [24] and the Cubic-Plus-Association (CPA) [25-27] are two of the most successful associating EoS. These are relevant to multicomponent, multiphase equilibria of systems containing associating components. A recent and successful EoS is the non-random hydrogen bonding (NRHB) [28,29], a comprehensive lattice model combined with an EoS theory, where holes are used to account for density variation as a result of temperature and pressure changes. This methodology accounts explicitly for the non-random distribution of molecular sites, being the model useful for property calculations of highly non-ideal fluids.

In this chapter, both activity coefficient models (NRTL-SAC, UNIFAC, A-UNIFAC) and an equation of state (the cubic-plus-association) were selected to represent the data. In the following sections, the most important equations and features of the selected models are briefly discussed. When required, the parameters obtained are presented. From what it is known, it was the first time that an equation of state was applied to predict phase equilibria of dihydroxibenzenes: until now, neither SAFT nor NRHB were applied with this purpose.

4.1. Nonrandom Two-Liquid Segment Activity Coefficient (NRTL-SAC) Model

The NRTL-SAC model is a quite recent and successful modification of the polymer NRTL equation, a derivative of the original NRTL model of Renon and Prausnitz [30], and has been widely applied to correlate and predict the phase equilibria of highly nonideal systems, like those containing complex organic electrolytes [31-34]. The fundamental idea is to limit the number of intermolecular interaction parameters needed and the essence of this model consists in representing the molecules by conceptual segments. Accordingly, the experimental measurements to be considered for parameter regression must be carefully selected to cover the full range of segments. One of the advantages of NRTL-SAC is that it only requires a well-chosen data to fit a smaller number of parameters. Accurate results were found in the solid-liquid equilibrium calculation of highly non-ideal systems [31-35].

The activity coefficient is obtained as a sum of two contributions: a combinatorialtermγ_I^C, calculated from the Flory-Huggins approximation for the combinatorial entropy of mixing; and a residualterm γ_I^R, set equal to the sum of the local composition (*lc*) interaction contribution for each segment. To account for several interactions, four predefined conceptual segments were proposed along with their corresponding segment-segment binary interaction parameters.This is the main difference between this model and the group-contribution approaches, which build molecules from a large set of predefined functional groups based on its chemical structure.

$$\ln \gamma_I^C = \ln \frac{\phi_I}{x_I} + 1 - r_I \sum_J \frac{\phi_J}{r_J} \tag{7}$$

with

$$r_I = \sum_m r_{m,I} \tag{8}$$

$$\phi_I = \frac{r_I x_I}{\sum_J r_J x_J} \tag{9}$$

where$J$ and I arecomponent indices, $m$represents a segment, $r_{m,I}$ is the number of segments of type m present in component I, r_I is the total segment number of component I, and $\phi_I$is the segment mole fraction of component I. The residual term is:

$$\ln \gamma_I^R = \sum_m r_{m,I} \left(\ln \Gamma_m^{lc} - \ln \Gamma_m^{lc,I} \right) \tag{10}$$

where Γ_m^{lc} is the activity coefficient of segment species m.

Each component is represented by one hydrophobic (X), one polar attractive (Y-), one polar repulsive (Y+), and one hydrophilic (Z) segment ($r_{m,I}$). The hydrophobic represents the molecular surface area unlike to form hydrogen bonds; the hydrophilic represents the area with interactions characteristic of a hydrogen bond donor or acceptor; and the polar segments represent interactions characteristics of an electron donor or acceptor.

To apply this model, an experimental database is used to estimate the solute model parameters, which afterwards can be adopted to predict the solubility in other solvent systems, at different temperatures. These conceptual segments are expected to describe the interaction characteristics of representative molecular surfaces, which significantly contribute to the liquid-phase non-ideality of real molecules.

In this essay, the hydroquinone parameters were estimatedusing solubility data in pure solvents, presented in Figure 12, melting data and the ΔC_pvalues (Eq. 5) and the solvent parameters published by Chen and Crafts [31]. Only six solvents were used to estimate four parameters, but these are considered to be enough to represent the molecular surface interaction characteristics. Esters are polar compounds with varying degrees of hydrophobic contents, alcohols are hybrids of hydrophobic and hydrophilic segments, and acids are complex molecules exhibiting hydrophilicity, polarity and hydrophobicity. It should be focused that water is not included in the database to regress the parameters, because in this way a higher predictive character is given to the model when the aqueous solubilities are estimated. The regressed parameters for hydroquinone are presented in Table 3.

Table 3. NRTL-SAC molecular parameters ($r_{m,I}$) for hydroquinone

X	Y-	Y+	Z
0.268	1.826	1.084	0.021

4.2. UNIFAC Method

As the experimental data available to calculate model parameters are scarce, group-contribution methods can be of great usefulness. In these methods, the molecule is divided into its functional groups, and the molecule-molecule interactions are considered to be weighed sums of group-group interactions: the main advantage is that the number of possible functional groups is much smaller than the number of distinct molecules. The UNIFAC method is one interesting approach, used, eg, in the estimation of solubilities in pure and mixed solvents. This is a predictive group-contribution method, which requires only chemical structure information for solutes and solvents. When not all group parameters are available, the method cannot be used. In these cases, the parametersmay be obtained from structurally similar molecules or from available experimental data. It often happens that the number of group parameters needing to be calculated is so high that the available database is insufficient to estimate them.

The activity coefficient is also obtained from the sum ofresidual and combinatorial contributions. This model was used here without any further modifications and as it is more popular than NRTL-SAC, its equations are not presented here, as theycan easily be found elsewhere [36]. The residual interaction parameters (a_{mn}) published by Hansen et al. [37] are used in this chapter, and characterize the interactions between groups$n$ and m. There are two parameters *per* group-group interaction ($a_{mn}\neq a_{nm}$), and no ternary parameters are required to describe multicomponent equilibria.

The dihydroxybenzenes are considered to be constituted by two –ACOH and four –ACH groups. The residual group interaction parameters used were those published by Hansen et al. [37], and the volume and area parameters were calculated according to the Gmehling et al. [38] revised tables.

4.3. A-UNIFAC Method

Sometimes UNIFAC shows limitations when strong association effects are present.In order to take this into account an extended version, the A-UNIFAC (association + UNIFAC), was developed by Mengarelli et al.[39]and applied to several complex systems[40-42].

According to this method, the activity coefficients are obtained by adding a third association term to the original model: this association term is derived from an expression to the residual Helmholtz energy based on a statistical-thermodynamic theory for molecules of classic fluids that tend to associate due to strongly directional association [43,44].

The association activity coefficient term $\gamma_I^{assoc.}$is a function of the fraction of non-bonded sites in the solution (X^{A_k}) and in pure-component I ($X_I^{A_k}$), calculated by:

$$\ln\gamma_I^{assoc} = \sum_{k=1}^{NAG}\left\{\vartheta_k^I\sum_{A_k}\left[\ln\left(\frac{X^{A_k}}{X_I^{A_k}}\right)+\frac{X_I^{A_k}-X^{A_k}}{2}\right]+\sum_{A_k}\left(\frac{1}{X^{A_k}}-\frac{1}{2}\right)N_k\left(\frac{\partial X^{A_k}}{\partial n_I}\right)_{T,P,n_j}\right\} \quad (11)$$

where N_k is the number of moles of associating groups$k$, represents the number of groups of type k in a molecule of component I, n_I is the number of moles of component I, NAG is the number of associating groups, and A_k is oneassociating site of group k. The fraction of non-bonded sites is a function of the associating strength $\left(\Delta^{A_k B_j}\right)$between site A of group k and site B of group j, and the associating group density in the mixture ρ_j[39,41]:

$$X^{Ak} = \frac{1}{1+\sum_{j=1}^{NAG}\sum_{Bj}\rho_j\Delta^{Ak,Bj}X^{Bj}} \qquad (12)$$

The fraction of non-bonded site Aofgroup $k$in the pure component I is a function of the density of group j in pure component I, $(\rho_j)_I$, and is given by:

$$X_I^{A_k} = \frac{1}{1+\sum_{j=1}^{NAG}\sum_{B_j}(\rho_j)_I\Delta^{A_k B_j}X_I^{B_j}} \qquad (13)$$

The association strength is a function of the energy and volume association parameters:

$$\Delta^{A_k B_j} = \kappa^{A_k B_j}\left[\exp\left(\frac{\varepsilon^{A_k B_j}}{kT}\right)-1\right] \qquad (14)$$

The final expression of the association contribution for the activity coefficient depends on the amount and type of the associating sites and species in solution, where the simplest expression corresponds to mixtures in which only one associating group is present. The dihydroxybenzenes have the two-site hydroxyl group, which can self- and cross-associate, and the one site aromatic ring, which can only cross-associate with electropositive sites. Solvents like acetic acid has the one-site carboxylic group, which can self- and cross-associate, and the esters, with one-site, can only cross-associate with electropositive associating sites.

The addition of the association term requires not only the estimation of the association parameters, but also a re-parameterization of the group interaction parameters in the residual term, because associative interactions are now explicitly taken into account.

The group assignments used to obtain the results presented in section 4.5werethe same as in the UNIFAC method, as well as the volume and area parameters. The residual group interaction parameters ($a_{m,n}$) as well as the A-UNIFAC association parameters used were those published by Ferreira et al.[41].

Hydroquinone has two associative hydroxyl groups that can self- and cross-associate; and the aromatic ring, which has one-site and can cross-associate with electropositive sites. For example, hydroquinone will cross-associate with water, alcohols and esters.

To apply this method to dihydroxybenzenes, it wasnecessary to extend the available residual group interaction parameter table: the parameters involving the –ACOH group had to be estimated in order to represent hydroquinone, catechol or resorcinol. The association parameters were taken from Ferreira et al. [41]and the association in this group is assumed to be the same as the association in the –OH group. These residual group interaction parameters were estimated using experimental data on low-pressure vapor-liquid equilibrium of binary mixtures (phenol in several solvents: benzene, alkanes, water, alcohols, esters and acids) and liquid-liquid equilibria of ternary mixtures (phenol + water + benzene; phenol + water + alcohols; phenol + water + ester; and phenol + water + acid). Table 4 reports the new group interaction parameters for –ACOH between paraffinic, aromatic, alcohol, water, acid and ester groups.

Table 4. Residual group interaction parameters, $a_{m,n}$ (K)

M	n	$a_{m,n}$	$a_{n,m}$
ACOH	CH_2	343.4	-55.40
	ACH	74.03	-42.87
	OH	-722.6	-410.1
	H_2O	-355.0	-179.2
	CCOO	-756.7	12.03
	COOH	402.9	1028

4.4. Cubic-Plus-Association Equation of State (CPA EoS)

The CPA EoS is the sum of two contributions: one accounting for physical interactions, like the Soave-Redlich-KwongEoS, and another accounting for the association interactions, originally proposed by Wertheim [43,44]. This last term takes into account the association due to hydrogen bonding between like molecules (self-association) and unlike molecules (cross-association orsolvation). The activity of each bonding site is assumed independent of the other bonding sites on the same molecule. In terms of Helmholtz energy (A), andusing a generalized cubic term,the CPAEoScan be written as:

$$A^{cubic} = \frac{an}{b(\delta_2 - \delta_1)} \ln\left(\frac{1 + b\rho\delta_1}{1 + b\rho\delta_2}\right) - nRT \ln(1 - b\rho) \quad (15)$$

$$A^{assoc} = RT \sum_i n_i \sum_{A_i} \left[\ln(X_{A_i}) - \frac{X_{A_i}}{2} + \frac{1}{2} \right] \quad (16)$$

where i is the component index, a is the energy parameter, b is the co-volume parameter, ρ the molar density, X_{A_i} the mole fraction of pure component i not bonded at site A, n is the total number of moles, n_i the number of moles of molecules of component i, and δ_1 and δ_2 are fixed to 1 and 0, respectively, whenever the SRK EoS is used in the cubic term. The pure component energy parameter is obtained from a Soave-type temperature dependency:

$$a(T) = a_0\left[1 + c_1\left(1 - \sqrt{T_r}\right)\right]^2 \quad (17)$$

where a_0 and c_1 are parameters to be estimated and T_r is the reduced temperature (the ratio between the absolute and the critical temperatures). The mixture a and$b$parameters are obtained through the classical mixing and combining rules[45]. Only a binary interaction parameter k_{ij} on the cross energy parameter a_{ij} is introduced in the cubic contribution [46].

X_{A_i} is related to the association strength $\Delta^{A_iB_j}$ between two sites belonging to two different molecules, and is calculated by:

$$X^{Ai} = \frac{1}{1+\rho \sum_j x_j \sum_{B_j} \Delta^{Ai,Bj} X^{B_j}} \quad (18)$$

For self-associating molecules:

$$\Delta^{A_iB_i} = g(\rho)\left[\exp\left(\frac{\varepsilon^{A_iB_i}}{RT}\right) - 1\right] b_i \beta^{A_iB_i} \quad (19)$$

where $\varepsilon^{A_iB_i}$ and are the association energy and the association volume, respectively. For sites that belong to two different associating molecules, the Elliot combining rule is used [47]:

$$\Delta^{A_iB_j} = \sqrt{\Delta^{A_iB_i} \times \Delta^{A_jB_j}} \tag{20}$$

The simplified radial distribution function $g(\rho)$ is given by[48]:

$$g(\rho) = \frac{1}{1-1.9\eta} \qquad \eta = \left(\frac{1}{4}\right)b\rho \tag{21}$$

being η the reduced fluid density.

The association term in CPA is characterized by the nature and number of associating groups, each of these being defined by an association scheme, as proposed by Huang and Radosz[49], as presented in Table 5.

Table 5. Association schemes [4

Species	Formula	Assigned scheme
Alcohols	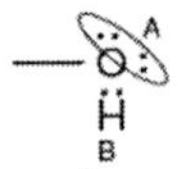	2B
Water	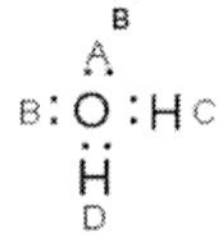	4C
Acids	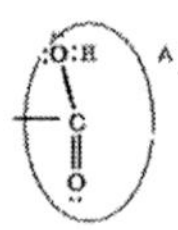	1A

The three cubic term parameters a_0, c_1 and b, and the associating parameters ε and $\beta$are the pure compound parameters of the model. These are usually optimized fitting pure component vapor pressure and liquid density data compiled from the DIPPR database [12], where the critical properties and the van der Waals volume, as well as property correlations as a function of temperature could also be collected. Regression of all the CPA parameters is performed in the reduced temperature range from 0.45 up to 0.85 where, in general, the DIPPR correlations are in agreement with the experimental data.

In general, the regression of a binary interaction parameter k_{ij} from experimental data is required to achieve a better representation for mixtures.

Still, there are many complex molecules for which pure component data are not available, and others for which a stable liquid does not exist. For hydroquinone, liquid density data are not available, and only the correlations could be used. Another major drawback is the presence of multiple associating groups on the same molecule. This motivated a search for a predictive scheme for obtaining the CPA parameters without requiring pure component data. In this predictive scheme, only the chemical structure of the molecule is required.It seems appropriate to individually assign association energies and volumes for each associating group and to consider multiple group substitution, either or not with the same association parameters. Thus, equation (19) will be used to obtain the association strengths (Δ) of the associating sites that belong to the same association group (with the same ε and β), whereas equation (20) will be used in the other cases, those either being self- or cross-association.

Parameters for the several solvents are well established and were collected from the literature [25,50] or, whenever unavailable, were estimated [51]. Resorcinol, catechol, hydroquinone, as well as other phenolic compounds, such as phenol, benzoic acid, salicylic acid, 2-phenylethanol and trans-cinnamic acid, were included in the database, being vapor pressure and liquid density data collected from the DIPPR database [12]. With this database, it was possible to compare the different cubic and associating CPA parameters and thus establish (1) correlations for the cubic parameters and (2) values for the association energies and volumes of the groups in different positions of the aromatic ring. Critical temperatures and pressures were obtained from a third-order group-contribution approach proposed by Marrero and Gani [52]. The van der Waals volume (*vdWV*) was obtained by the Bondi's group contribution approach [53]. The experimental [12] and estimated values are shown in Table 6. The critical pressures and van der Waals volumes are in very good agreement, as well as the critical temperature of catechol. For the critical temperatures of resorcinol and hydroquinone, slightly higher deviations were found. The critical temperatures together with the available vapor pressure and liquid density data were used to correlate simultaneously all CPA parameters.

The obtained cubic parameters in this step were then regressed: a_0 and c_1 as a function of T_c^2/P_c and T_c, respectively, while the b parameter wasrelated to the van der Waals volume obtained from the Bondi's group contribution approach.

Table 6. Critical properties and van der Waals volume

Compound	T_c (K)		P_c (MPa)		$vdWV$ x 10^5 (m^3/mol)	
	Exp.	Pred.	Exp.	Pred.	Exp.	Pred.
Catechol	764.0	757.9	7.49	7.43	5.94	5.98
Resorcinol	810.0	760.1	7.49	7.29	5.94	5.98
Hydroquinone	823.0	761.4	7.45	7.44	5.94	5.98
Phenol	694.3	687.0	6.13	5.96	5.38	5.28

The following equations represent the best linear fit of the cubic parameters [54,55]:

$$a_0 = 0.2267 + 24.38\left(\frac{T_c^{\,2}}{P_c}\right) \tag{22}$$

$$c_1 = -3.557 + 6.289 \times 10^{-3} T_c \tag{23}$$

$$b = -2.328 \times 10^{-5} + 1.884\, vdWV \tag{24}$$

whereT_c is given in Kelvin, P_c in Pascaland *vdWV* is expressed in m^3/mol.

The cubic parameters were then fixed by these equations, and the associating parameters regressed separately. The obtained parameters are presented in Table 7.

Table 7. Associating pure-component parameters, and deviations in the estimation of vapor pressure (P) and liquid density (ρ)using the cubic term parameters from Eqs. 21-23. (For these molecules multiple hydroxyl groups within the same molecule are considered to have the same association parameters)

Compound	OH		AAD (%)	
	ε (kJ/mol)	β	P	ρ
Catechol	11.3	6.23 x 10^{-2}	4.77	2.23
Resorcinol	18.5	2.21 x 10^{-2}	0.62	4.13
Hydroquinone	18.7	2.66 x 10^{-2}	1.66	20.1
Phenol	18.4	1.19 x 10^{-2}	0.90	6.94

It can be seen that, with the exception of catechol where intramolecular hydrogen bonding occurs, very close association energies were obtained. Whenever intramolecular hydrogen bonding occurs this value is slightly lower, as expected. The association volume values are more dependent on the substitution of the ring. Discrepancies in the association volume are more expectable since it depends more strongly on the overall molecular structure, while the association energy is related to the localized hydrogen bonding formation.

4.5. Results and Discussion

When using activity coefficient models, the vapor phase is regularly assumed to be ideal. In this case, solution models including the Wilson, NRTL, and UNIQUAC can be utilized to represent the non-ideality of the liquid phase. All of these three models correlated equally well for 4-methoxyphenol+catechol and *p*-cresol+4-methoxyphenol. Better results wereobtained usingthe NRTLmodel for *p*-cresol+catechol. The results for 4-methoxyphenol + catechol system exhibit strong negative deviations from Raoult's law, most likely due to hydrogen bonding between these two dissimilar molecules. But, the Wilson model is perfectly able to correlate such VLE data. The bubble pressures and saturatedvapor compositions for the ternary system were predicted using directly the binary parameters calculatedfromthe binary VLE data. The NRTL model gives the best predictions.In spite ofthe deviationsbeingslightly largerthan for the binary systems, comparable results are obtained for the ternary system [13]. Once again, the modeling results reinforce the difficulty to separate the mixture 4-methoxyphenol + catechol using conventional distillation, due to the existence of a minimum pressure azeotrope.

Yau and Tsal [14]showed that the solubility of carbon dioxide in phenol and catechol may be represented adequately by the modified Soave equation with one interaction parameter. They have also determined the Henry's constant and partial molar volume at infinite dilution by fitting the data, as well as the Margules parameter.If a comparison between calculated and experimental solubilities of carbon dioxide was done, it can be seen that the results are in agreement over the entire temperature and pressure ranges.

Yamini et al.[15] correlated the solubility of dihydroxybenzene isomers in supercritical carbon dioxide using the Chrastilmodel, which relatesthe solubility to the density of the supercritical fluid, based on the assumption that the association of a solute (S) with k molecules of supercritical solvent (C) produces a solvato complex, which is in equilibrium with gas:

$$\ln s = k \ln \rho + \frac{a}{T} + b \tag{25}$$

where

$$a = \frac{\Delta H}{R} \qquad and \qquad b = -\ln\left[\frac{M_C^k}{M_S + kM_C}\right] + q \tag{26}$$

s is solubility, ρ is the solvent density, k is the association number, ΔH is thetotal heat of dissolution(i.e., heat of solvation plus heat of vaporization of the solute), q is a constantand,$M_S$and $M_C$are the molecular weights of solute and solvent, respectively. The associationconstant k is the slope of the linear correlation ln$s$ vs. 1/T and represents the average number of solventmolecules in the solvato complex. lnsis a linear function of $1/T$ at constant density and has aslope given by a. This model assumes the density as temperature independent which in fact is a simplification because it is temperature dependent.The value of the constant b can be chosen to minimize the deviation of the model fromexperimental data. The data obtained in the referred study show the expected linear relationship predicted by the Chrastilmodel. The equilibrium solubilities of dihydroxybenzene isomers pyrocatechol, resorcinol and hydroquinone in supercritical carbon dioxide were estimated, in the pressure range from 120 to 400 atm, at temperatures 308, 318, 328 and 338 K.The slopes of the solubility isotherms are with a good approximation equal. The average absolute relative deviations in gas phase composition are within 3%.

Experimental data obtained by Coutsikos et al. [16]for the solubility of various phenolics in carbon dioxide were correlated with the translated and modified Peng-Robinson cubic equation of state. The results were of good quality and could be used for the design of extraction processes for these compounds: deviations between 11 and 19% were obtained.

The experimental solubilities measured by García-González et al. [17] for binary and ternary systems of hydroquinone and p-quinone in supercritical carbon dioxide, were correlated using a semi-empirical Bartle model, which relates the mole fraction solubility with pressure and density of the solution. A standard pressure and density should be defined, to make the constant values much less sensitive to the experimental error and to avoid large deviations caused by extrapolations to zero density. The data obtained in thatstudy showed the expected linear relationship predicted by the Bartle model ($\ln\left(\frac{yP}{P_{ref}}\right)$ vs. ρ) for hydroquinone and p-quinone both in the binary and ternary systems. Agreement between the calculated and experimental values for both compounds in the binary and ternary systems wassatisfactory.The Bartle method provided a good fit, with average absolute relative deviations lower or equal to9.6% in all cases.

In the following,a comparison of the solubilities estimation of hydroquinone and its isomers in different solvents using the more recent thermodynamic models previously described insections 4.1-4.4, is presented. In order to compare the experimental solubility values taken from literature with the calculated results, absolute average deviations (AAD) are used:

$$AAD\ (\%) = \frac{1}{NP}\frac{\left|x^{\exp} - x^{calc}\right|}{x^{\exp}} \times 100 \quad (27)$$

where NP is the number of data points, x is the solubility, and the superscripts *exp* and *calc* refer to experimental and calculated values, respectively. Table 8shows the AAD's for the solubility calculations in pure (liquid) solvents using the excess Gibbs energy models and the CPA equation of state, together with the average AAD for each model. It is important to mention that when the AAD is higher than 150%, this error is excluded from the average presented for each model, in order that the averages are not biased.

Within the excess Gibbs energy models, the results are better with UNIFAC and NRTL-SAC, except in water where, in spite of the deviation being still high, a great improvement was achieved using A-UNIFAC. This proofs the importance of taking into account the association effects. This is clearly showed by the good results achieved also with the CPA EoS, using a single temperature-independent k_{ij} for each binary system.

For the NRTL-SAC model, the deviations found in water and acetic acid are exceptionally high, but it should be mentioned once again that water was not included in the database for the regression of the parameters, and acetic acid is an outlier in the correlation results found for NRTL-SAC. So, these worst results for these two solvents weremore or less expected. The higher deviation found in methanol using A-UNIFAC, when compared to the other alcohols, was probably due to the group assignment considered for methanol: this was not the most correct.

Table 8. AAD's for solubility calculationsof hydroquinone in pure solvents

Solvent	A-UNIFAC	UNIFAC	NRTL-SAC	CPA [51,56]
water	116	602	600	43.2
methanol	92.2	39.0	18.6	6.27
ethanol	21.1	11.1	29.4	22.6
2-propanol	30.4	25.4	24.6	8.63
ethyl acetate	118	91.2	44.5	23.0
butyl acetate	62.4	34.2	42.7	22.2
acetic acid	82.1	66.6	200	8.32
	75.0	44.6[1]	32.0[1]	19.2

1Without outliers: AAD > 150%.

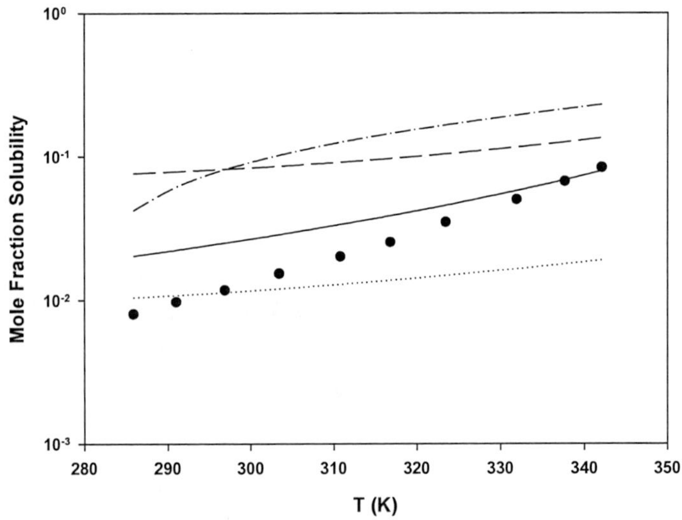

Figure 15. Aqueous solubilities of hydroquinone [18] as a function of temperature and modeling results (–, A-UNIFAC; --, UNIFAC; ···, CPA; -·-, NRTL-SAC).

Until now, there are no parameters for the methanol group in the A-UNIFAC group-interaction tables [41], as in UNIFAC tables, so a CH3 + OH had to be considered to represent methanol for the A-UNIFAC method. At this point, it should be referred that the estimation of the –ACOH group interaction parameters for the A-UNIFAC method had a great importance in the improvement of the results: if the group assignment was considered to be –AC + –OH instead, the method showed special difficulties to deal with aromatic alcohols, presenting deviations around 90%. Figure 15 gives the comparison results between models for the aqueous solubilities of hydroquinone as a function of temperature. Concerning the CPA EoS, good results were achieved but at the cost of more pronounced corrections (-0.13 $<k_{ij}<$ -0.23), except for acetic acid where the predictions are remarkably good. The solubilitiesin this solvent are the lowest, although in the order of 10^{-2} and, like that, does not contradict some indications that larger deviations are found in solvents where the solubilities are very low. All the correlated k_{ij} are negative, indicating that the solvent-solute interactions are stronger than expected by the model. This is particularly the case of the esters (-0.196 and -0.231, respectively for ethyl and butyl acetate), where the presence of the ester group was not considered to increase solvation.

In all systems, CPA results are considerably better, except in ethanol where all the models work reasonably well, what is very good given the higher predictive character of CPA compared to NRTL-SAC. In Figure 16, a comparison between the excess Gibbs energy models and CPA results for methanol and the other organic solvents (Table 8) are presented.

Additionally, it was found that the CPA EoSis perfectly able to represent the solid-liquid equilibria of phenol in water (deviation lower than 13%), where the group-contribution methods give results completely out of range, as can be seen in Figure 17.

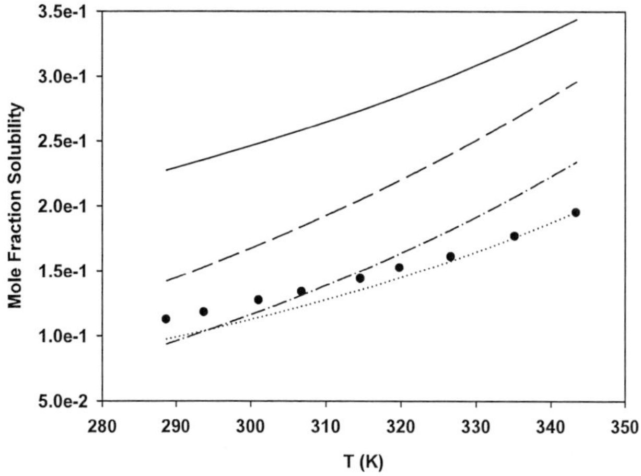

(a)

Figure 16. (Continued).

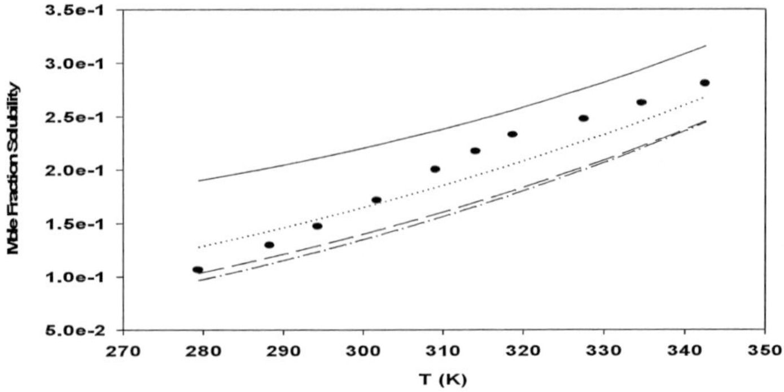

(b)

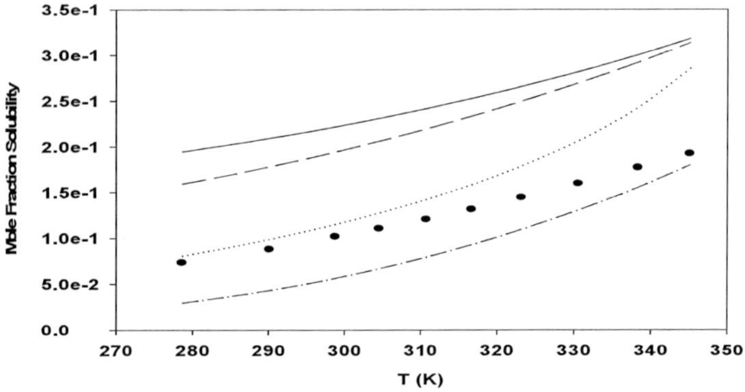

(c)

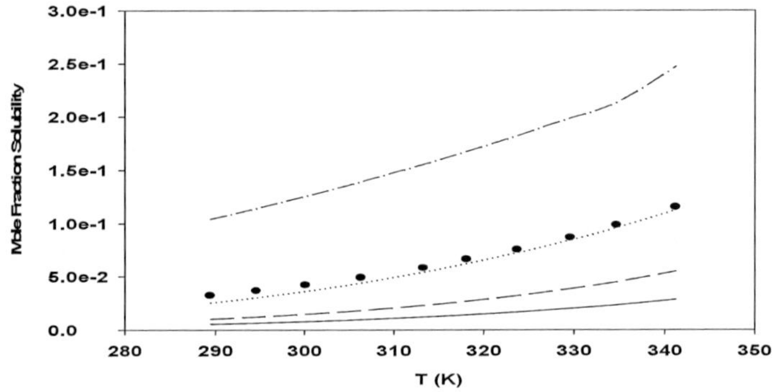

(d)

Figure 16. Solubilities of hydroquinone as a function of temperature in (a) methanol; (b) 2-propanol; (c) ethyl acetate; and (d) acetic acid: experimental and modeling results (–, A-UNIFAC; --, UNIFAC; ···, CPA; -··-, NRTL-SAC).

It is expected that A-UNIFAC works much better for this system since the -ACOH group interaction parameters were estimated based on vapor-liquid and liquid-liquid equilibrium data of phenol systems, and for these cases the correlation results were very accurate. However, when addressing solid-liquid equilibria the method presented difficulties in founding the estimates.

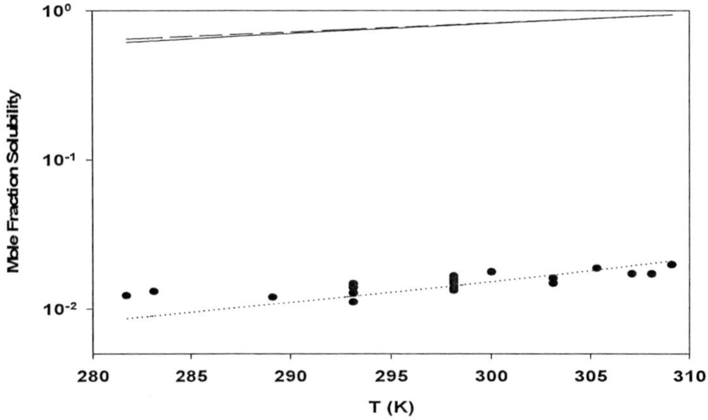

Figure 17. Solubilities of phenol as a function of temperature in water: experimental [18] and modeling results (–, A-UNIFAC; --, UNIFAC; ···, CPA).

For the equilibria of catechol and resorcinol in water, the available data are muchdispersed. In this way, van't Hoff plots (ln x vs. 1/T) were analyzed and the experimental points with considerable differences to the empirical correlation curve were considered outliers, and rejected.

In the case of CPA EoS, solubility data are then estimated from these correlations and used to estimate the k_{ij}allowing a smoothed procedure.The best model to represent their solubility data is preciselythe CPA EoS with global deviations in solubility of 77 and 59%, respectively. Still, in those cases, even if only a single and temperature-independent k_{ij} value was required for the CPA EoS, this correction parameter was large (around -0.1).

A-UNIFAC leads to a great improvement when compared to UNIFAC, which, once again, proves the importance of taking association effects into account.

In Figure 18, the experimental and calculated solubilities in water are compared. The data rejected are also included enabling a better understanding about the difficulties to determine these solubilities not only when modeling, but also in terms of experimental measurements. The NRTL-SAC could not be applied for these compounds because the solute molecular parameters are not available and cannot be regressed since solubility data in at least five solvents are not available as well.

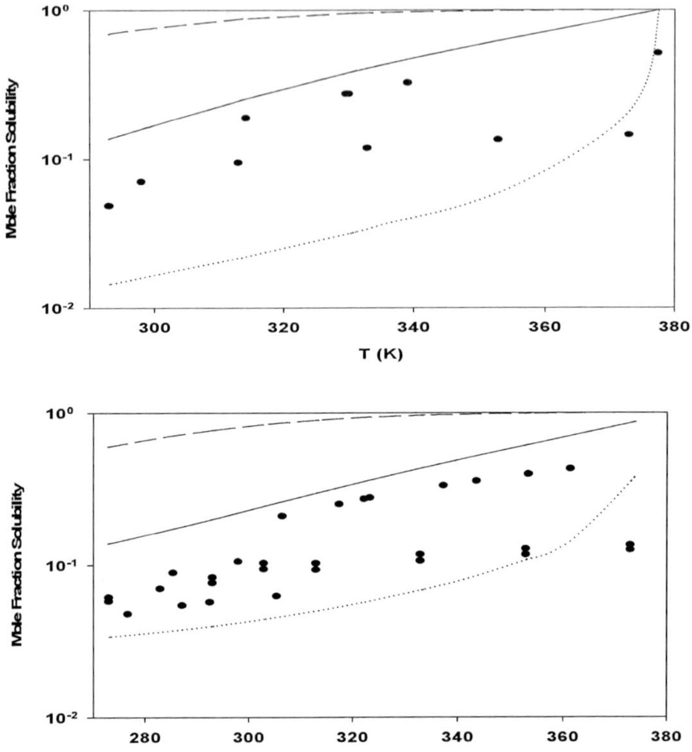

(a) (b)

Figure 18. Solubilities of (a) catechol and (b) resorcinol as a function of temperature in water: experimental [18] and modeling results (–, A-UNIFAC; --, UNIFAC; ···, CPA).

Conclusion

The benzenediols hydroquinone, catechol and resorcinol, are a family of phenolic compounds, with two hydroxyl groups attached to an aromatic ring. They have well known applications in the food andpharmaceuticalindustries. In this chapter, the principal routes of manufacture and processing were reviewed.

Concerning the purification and separation processes, distillation is not a reliable alternative to separate benzenediol mixtures due to their similar vapor pressures. Thus, the use of distillation in the preparation of benzenediols is limited to solvent recovery.

Crystallization is another important separation and purification process employed to produce a broad range of materials, and usually, cooling crystallization is used to purify these compounds. To select the appropriate solvents for crystallization, an adequate knowledge of phase equilibria is required.

Supercritical fluid extraction using carbon dioxide has important advantages over conventional extraction techniques: the absence of both light and air during the extraction process, along with the low operating temperatures reduce the incidence of degradation reactions, while the use of carbon dioxide as extracting agent allows working in an environmentally clean media.

In this chapter, some phase equilibria data are reviewed, as well as it is provided a comparison of different modeling methodologies to estimate the solubility of dihydroxybenzenes in aqueous, organic and supercritical media.

In this way, vapor-liquid, supercritical and solid-liquid equilibriadata were presented for catechol, resorcinol and hydroquinone, and for comparison purposes some data for phenol.

For the first time, the group-contribution methods, UNIFAC and A-UNIFAC, were compared with the NRTL-SAC and the CPA equation of state, to predict the solubility in pure solvents as a function of temperature. It was proved the importance of the A-UNIFAC and CPA EoS for describing associative systems, as the compounds under study are associative compounds as well as the majority of solvents.

For aqueous systems, A-UNIFAC was in general superior to UNIFAC, proving the need of taking into account the association effects.

The NRTL-SAC model was successfully applied to estimate the hydroquinone solubilities, except in water and acetic acid. Solubility in pure solvents was used to regress the solute model parameters, which were used afterwards for the prediction of the solubility of these compounds in water and for correlation as a function of temperaturefor other solvents.

The Cubic-plus-Association equation of state (CPA EoS) was used considering explicitly all associating groups in the molecule. A methodology to obtain the pure-component parameters solely from their molecular structure without requiring experimental data was used, and the results indicate that CPA is an adequate tool to correlate the water and organic solvent solubilitiesof dihydroxybenzenes. The methodology proposed is expected to become an important tool for the modeling of the phase equilibria of such molecules.

REFERENCES

[1] Olumide, Y. M.; Akinkugbe, A. O.; Altraide, D.; Mohammed, T.; Ahamefule, N.; Ayanlowo, S.; Onyekonwu, C.; Essen, N.; *Int. J. Dermatol.* 2008, 47, 344.

[2] Li, X. N.; Yin, Q. X.; Chen, W.; Wang, J. K.; *J. Chem. Eng. Data* 2006, 51, 127.

[3] DeCaprio, A. P.; *Crit. Rev. Toxicol.* 1999, 29, 283.

[4] Suda, H.; Dohgane, I.; Chinuki, T.; Tanimoto, K.; Hosaka, H.; Ebara, K.; Nakao, Y.; Ueda, Y.; Imada, S.; Yasuda, M.; Process for the separation of resorcinol and hydroquinone from their admixtureed.; United States, 1975.3911030.

[5] Suda, H.; Dohgane, I.; Chinuki, T.; Tanimoto, K.; Hosaka, H.; Nakao, Y.; Ueda, Y.; Imada, S.; Yanagihara, H.; Tanaka, K.; Process for separating resorcinol from hydroquinoneed.; United States, 1975.3929910.

[6] Freitag, D.; Geyer, H.; Kraus, A.; Viswanathan, R.; Kotzias, D.; Attar, A.; Klein, W.; Korte, F.; *Ecotoxicol. Environ. Saf.* 1982, 6, 60.

[7] Geyer, H.; Viswanathan, R.; Freitag, D.; Korte, F.; *Chemosphere* 1981, 10, 1307.

[8] Zheng, L. T.; Ryu, G. M.; Kwon, B. M.; Lee, W. H.; Suk, K.; *Eur. J. Pharmacol.* 2008, 588, 106.

[9] Hahn, S.; Kielhorn, J.; Koppenhöfer, J.; Wibbertmann, A.; Mangelsdorf, I.; Resorcinol. In Data, W. L. C.-i.-P. (eds).World Health Organization, Stuttgart (2008).

[10] Krumenacker, L.; Costantini, M.; Pontal, P.; Sentenac, J. *Hydroquinone, Resorcinol, and Catechol*; John Wiley & Sons, Inc., 2000.

[11] Mullin, J. W. *Crystallization*; Butterworth-Heinemann: Oxford, 1993.

[12] *DIPPR Tables*; *Brigham Young University: Provo, UT* 1998.

[13] Hwang, S. M.; Lee, M. J.; Lin, H. M.; *Fluid Phase Equilib.* 2000, 172, 183.

[14] Yau, J. S.; Tsai, F. N.; *J. Chem. Eng. Data* 1992, 37, 141.

[15] Yamini, Y.; Fat'hi, M. R.; Alizadeh, N.; Shamsipur, M.; *Fluid Phase Equilib.*1998, 152, 299.

[16] Coutsikos, P.; Magoulas, K.; Tassios, D.; *J. Chem. Eng. Data* 1995, 40, 953.

[17] Garcia-Gonzalez, J.; Molina, M. J.; Rodriguez, F.; Mirada, F.; *Fluid Phase Equilib.*2002, 200, 31.

[18] Yalkowsky, S. H.; He, Y. *Handbook of Aqueous Solubility Data*; CRC Press: New York, 2003.

[19] Adkins, C. J. *Equilibrium Thermodynamics*; Cambridge University Press: New York, 1983.

[20] Prausnitz, J. M.; Lichtenthaler, R. N.; Azevedo, E. G. *Molecular Thermodynamics of Fluid-Phase Equilibria*; Prentice Hall PTR: New Jersey, 1999.

[21] Huuskonen, J.; *Comb. Chem. High Through. Screen.* 2001, 4, 311.

[22] Jorgensen, W. L.; Duffy, E. M.; *Adv. Drug Del. Rev.* 2002, 54, 355.

[23] Millard, J. W.; Alvarez-Nunez, F. A.; Yalkowsky, S. H.; *Int. J. Pharm.* 2002, 245, 153.

[24] Economou, I. G.; *Ind. Eng. Chem. Res.* 2002, 41, 953.

[25] Kontogeorgis, G. M.; Michelsen, M. L.; Folas, G. K.; Derawi, S.; von Solms, N.; Stenby, E. H.; *Ind. Eng. Chem. Res.* 2006, 45, 4869.

[26] Kontogeorgis, G. M.; Michelsen, M. L.; Folas, G. K.; Derawi, S.; von Solms, N.; Stenby, E. H.; *Ind. Eng. Chem. Res.* 2006, 45, 4855.

[27] Kontogeorgis, G. M.; Voutsas, E. C.; Yakoumis, I. V.; Tassios, D. P.; *Ind. Eng. Chem. Res.* 1996, 35, 4310.

[28] Panayiotou, C.; Tsivintzelis, I.; Economou, I. G.; *Ind. Eng. Chem. Res.* 2007, 46, 2628.

[29] Tsivintzelis, I.; Spyriouni, T.; Economou, I. G.; *Fluid Phase Equilib.*2007, 253, 19.

[30] Renon, H.; Prausnitz, J.; *AlChE J.* 1968, 14, 135.

[31] Chen, C. C.; Crafts, P. A.; *Ind. Eng. Chem. Res.* 2006, 45, 4816.

[32] Chen, C. C.; Song, Y. H.; *Ind. Eng. Chem. Res.* 2004, 43, 8354.

[33] Kokitkar, P. B.; Plocharczyk, E.; *Org. Proc. Res. Dev.* 2008, 12, 249.

[34] Mota, F. L.; Carneiro, A. P.; Queimada, A. J.; Pinho, S. P.; Macedo, E. A.; *Eur. J. Pharm. Sci.* 2009, 37, 499.

[35] Tung, H. H.; Tabora, J.; Variankaval, N.; Bakken, D.; Chen, C. C.; *J. Pharm. Sci.* 2008, 97, 1813.

[36] Fredenslund, A.; Gmehling, J.; Michelsen, M. L.; Rasmussen, P.; Prausnitz, J. M.; *Ind. Eng. Chem. Proc. Des. Dev.* 1977, 16, 450.

[37] Hansen, H. K.; Rasmussen, P.; Fredenslund, A.; Schiller, M.; Gmehling, J.; *Ind. Eng. Chem. Res.* 1991, 30, 2352.

[38] Gmehling, J.; Li, J. D.; Schiller, M.; *Ind. Eng. Chem. Res.* 1993, 32, 178.

[39] Mengarelli, A. C.; Brignole, E. A.; Bottini, S. B.; *Fluid Phase Equilib.* 1999, 163, 195.

[40] Ferreira, O.; Brignole, E. A.; Macedo, E. A.; *Ind. Eng. Chem. Res.* 2003, 42, 6212.

[41] Ferreira, O.; Macedo, E. A.; Bottini, S. B.; *Fluid Phase Equilib.* 2005, 227, 165.

[42] Montañés, F.; Olano, A.; Ibáñez, E.; Fornari, T.; *AlChE J.* 2007, 53, 2411.

[43] Wertheim, M. S.; *J Stat Phys* 1984, 35, 19.

[44] Wertheim, M. S.; *J Stat Phys* 1984, 35, 35.

[45] Oliveira, M. B.; Coutinho, J. A. P.; Queimada, A. J.; *Fluid Phase Equilib.* 2007, 258, 58.

[46] Oliveira, M. B.; Freire, M. G.; Marrucho, I. M.; Kontogeorgis, G. M.; Queimada, A. J.; Coutinho, J. A. P.; *Ind. Eng. Chem. Res.* 2007, 46, 1415.

[47] Voutsas, E. C.; Yakoumis, I. V.; Tassios, D. P.; *Fluid Phase Equilib.*1999, 158, 151.

[48] Kontogeorgis, G. M.; Yakoumis, I. V.; Meijer, H.; Hendriks, E.; Moorwood, T.; *Fluid Phase Equilib.* 1999, 160, 201.

[49] Huang, S. H.; Radosz, M.; *Ind. Eng. Chem. Res.* 1990, 29, 2284.

[50] Folas, G. K.; Kontogeorgis, G. M.; Michelsen, M. L.; Stenby, E. H.; *Ind. Eng. Chem. Res.* 2006, 45, 1527.

[51] Mota, F. L.; Queimada, A. J.; Pinho, S. P.; Macedo, E. A.; *Fluid Phase Equilib.* 2011, 303, 62.

[52] Marrero, J.; Gani, R.; *Fluid Phase Equilib.* 2001, 183, 183.

[53] Bondi, A.; *J. Phys. Chem.* 1964, 68, 441.

[54] Mota, F. L.; Queimada, A. J.; Pinho, S. P.; Macedo, E. A.; *Ind. Eng. Chem. Res.* 2008, 47, 5182.

[55] Queimada, A. J.; Mota, F. L.; Pinho, S. P.; Macedo, E. A.; *J. Phys. Chem. B* 2009, 113, 3469.

[56] Mota, F. L.; Queimada, A. J.; Pinho, S. P.; Macedo, E. A.; *Fluid Phase Equilib.*2010, 298, 75.

In: Hydroquinone
Eds: F. Gokden and A. Lazzarotto
ISBN: 978-1-62100-258-1

Chapter 3

Permeation and Redox Mediation Processes at Poly(*o*-Aminophenol) Film Electrodes in the Presence of Hydroquinone/*p*-Benzoquinone Redox Species: A Review Article

***R. Tucceri*[1,*], *P. M. Arnal*[2] *and A. N. Scian*[3]**
[1]Instituto de Investigaciones Fisicoquímicas Teóricas y Aplicadas (INIFTA), Facultad de Ciencias Exactas, Universidad Nacional de La Plata, La Plata, Argentina
[2]Centro de Tecnología de Recursos Minerales y Cerámica (CETMIC), CIC - CONICET La Plata, Buenos Aires, Argentina
[3]Centro de Tecnología de Recursos Minerales y Cerámica (CETMIC), CIC - CONICET La Plata, Buenos Aires, Argentina

Abstract

This review paper describes the different researches carried out in our laboratory to study transport and transfer processes at poly(*o*-aminophenol) (POAP) film electrodes in the presence of the

---

* Corresponding author: Mailing Address: R.I. Tucceri, Instituto de Investigaciones Fisicoquímicas Teóricas y Aplicadas (INIFTA), Sucursal 4, Casilla de Correo 16, (1900)-La Plata, Argentina. Fax: (54) (0221) 425-4642. E-mail: rtucce@inifta.unlp.edu.ar.

hydroquinone/*p*-benzoquinone (HQ/Q) redox couple. The most significant experiments employing different techniques such as Cyclic Voltammetry, Rotating Disc Electrode Voltammetry and Electrochemical Impedance Spectroscopy are described in detail, and the application of existing models and theories that allow extracting charge-transport and charge-transfer parameters are also outlined. While an electron-transfer reaction (mediation reaction) occurs at negative potential values ($E < 0.0$ V *vs.* SCE), permeation processes are observed at positive potential values ($E > 0.8$ V *vs.* SCE). Freshly prepared poly(*o*-aminophenol) films were quantitatively deactivated to study how their permeation and charge-transfer processes are affected by the degree of deactivation (θ_c). While for low degrees of deactivation ($\theta_c < 0.4$) it is possible to recover the conductivity of a deactivated POAP film by reactivation in an alkaline solution, high degrees of deactivation lead to an irreversible deterioration of the polymer film. Electrochemical Impedance Spectroscopy applied at negative potential values allowed obtaining dependences of the different transport parameters on the degree of deactivation of the polymer film. While some parameters such as interfacial metal-film and film-solution resistances ($R_{m|f}$, $R_e^{f|s}$, $R_i^{f|s}$), the high-frequency capacitance (C_H) and the redox capacitance (C_p) exhibit a continuous variation without hysteresis between deactivation and reactivation processes within the whole θ_c range, others such as electron and ion diffusion coefficients (D_e, D_i) show not only marked changes of slope from given θ_c values but also hysteresis between consecutive deactivation and reactivation processes. The diffusion rates of the hydroquinone and benzoquinone species across the polymer film are also strongly reduced with the increase in the degree of degradation. This work could be interesting due to the wide range of potential applications of POAP. In this sense, although in practical applications of POAP it is necessary to maintain the conducting properties unaltered, the polymer is subjected to extreme conditions that can cause its partial deactivation. Thus, it seems to be important for electrochemists to know at least how the charge-transport process at POAP films changes with their deactivation.

Keywords: poly(*o*-aminophenol) film electrodes; deactivation and reactivation processes; *p*-benzoquinone-hydroquinone species; charge-transport and charge-transfer parameters

1. INTRODUCTION

Traditionally polymers were thought of as insulators and any electrical conduction in polymers was generally regarded as an undesirable phenomenon. The residual conductivity in common polymers can mostly be assigned to loosely bound protons. In the last decades an opposite trend has started inasmuch as investigations have been directed to the utilization of ionic conductivity of polymeric systems. The active research on thermodynamic and kinetic properties of ion-conducting polymers has led to the wide use of polymer electrolytes and polyelectrolytes in electrochemical systems, e.g. in power sources, sensors, and development of all-solid-state electrochemical devices [1-3]. Also in the 1970s somewhat surprisingly a new class of polymers possessing high electronic conductivity (electronically conducting polymers) in partially oxidized (or less frequently reduced) state was discovered. Electrochemistry has played a significant role in the preparation and characterization of these novel materials. Electrochemical techniques are especially suitable for controlled synthesis of these compounds and for tuning of a well-defined oxidation state. The preparation, characterization and applications of electrochemically active, electronically conducting polymeric systems are still in the foreground of research activity in electrochemistry. There are at least two major reasons for this intense interest. First, the intellectual curiosity of scientists that focuses on understanding the behaviour of these systems, in particular on the mechanism of charge-transfer and charge-transport processes occurring in the course of redox reactions of conducting polymeric materials. Second, the wide range of promising applications in the field of energy storage, electrocatalysis, organic electrochemistry, bioelectrochemistry, photoelectrochemistry, electroanalysis, sensors, electrochromic displays, microwave screening and corrosion protection. Several excellent monographs have been published, e.g. [4-6] mostly reviewing the knowledge accumulated regarding the development of polymer film electrodes and their applications.

Electroactive polymers contain electron donor and acceptor sites that usually consist of up to hundreds of molecular layers [7-9]. Oxidation and reduction of fixed sites introduce charged sites into the polymer, which, to achieve charge neutrality, require the ingress of counter-ions from the contacting electrolyte solution and, according to the Donnan relation, the egress of co-ions [10]. Electron hopping is believed to be the mechanism for electron transport, but it is also possible that ion motions may partially or totally control the rate of charge transport. Thus, the charge-transport rate in electroactive polymer films is quantitatively characterised by a heterogeneous constant of charge transfer and an effective (or apparent) diffusion coefficient for charge transport across the film [11, 12]. The factors governing the magnitude of these parameters are not well understood. Furthermore, in general, these transport parameters depend on different variables such as degree of oxidation of the polymer, polymer film thickness, type of electrolyte, ionic strength and *pH* of the solution in contact with the polymer film. A combination of different electrochemical techniques was employed to determine the rates of the charge-transfer and charge-transport processes in electroactive polymers, such as Cyclic Voltammetry (CV), Rotating Disc Electrode Voltammetry (RDEV) and Electrochemical Impedance Spectroscopy (EIS) [12]. When redox species are present in the electrolyte solution, they can participate in an interfacial electron exchange with the polymer at the polymer|solution boundary [13-16]. In this regard, an electroactive polymer can act as a redox mediator. Redox mediators are substances containing redox centres that may be oxidized and reduced more or less reversibly. In general, redox mediation is a phenomenon by which a substance of low conductivity in contact with a metal, on one side, and in contact with a solution containing one or both components of a redox couple, on the other, is capable of mediating electron transfer between the redox couple in solution and the metal. In the context of the RDEV and the EIS, the conduction properties of electroactive polymer films in the presence of redox species are often tested by analysing the electrochemical behaviour of the films as a function of the electrode rotation rate, polymer film thickness and redox couple concentration.

Poly-*o*-aminophenol (POAP) is an electroactive polymer, which has been synthesized on different electrode materials [17-20] and characterized by using different electrochemical techniques [21-31]. Several practical applications of POAP in the field of biosensors [32-39] and electrocatalysis [40-42] have been reported in the recent literature. POAP synthesized in acidic media has been used as a potentiometric ferric cation sensor in solution [43] and in the development of a sensor to determine the concentration of NO in solution [44]. Also, POAP synthesized in neutral media was employed as component in sensors of glucose [40], H_2O_2 [39] and uric acid [36]. From the practical viewpoint it is necessary that the conducting properties of POAP remain substantially unchanged, before and after the polymer has been used. Then, the study of the charge conduction process at this polymer is a matter of considerable interest.

This review mainly concerns charge-transport and charge-transfer processes at poly(*o*-aminophenol) (POAP) film electrodes in the interesting case where the polymer contacts a solution containing the hydroquinone/*p*-benzoquinone (HQ/Q) redox couple. In this regard, we have attempted to document the most significant evidence, which proves that the electrochemical behaviour of POAP in contact with the HQ/Q redox couple can be explained on the basis of a rapid electron-transfer process at the POAP|solution interface at negative potential values ($E < 0.0$ V *vs.* SCE) and a permeation process across the polymer film at positive potential values ($E > 0.8$ V *vs.* SCE). Several electrochemical techniques such as Cyclic Voltammetry (CV), Rotating Disc Electrode Voltammetry (RDEV) and Electrochemical Impedance Spectroscopy (EIS) were employed in this study to analyze how the electron conduction process and the physical diffusion rate of a neutral species across the polymer film are affected by deactivation and degradation of the polymer.

This review could be interesting not only to the electrochemists who have worked for years in basic research related to the conducting properties of this polymer, but also to readers interested in biosensors based on POAP for clinical and industrial applications, where the polymer operates under limit conditions, which can affect its conductivity and then, reactivation of the polymer becomes necessary.

2. Transport Processes at Poly(*O*-Aminophenol) (POAP) Films in the Presence of the Neutral HQ/Q Redox Couple. Studies Applying Cyclic Voltammetry (CV) and Rotating Disc Electrode Voltammetry (RDEV)

2.1. Electroactive POAP Films in Contact with a Solution Containing the HQ/Q Redox Couple

The first study about the conduction processes at POAP film electrodes in the presence of the hydroquinone electroactive species was reported in [18]. By employing CV it was demonstrated that chemical substances that have a redox potential ($E^{o'}$) value more positive than the anodic peak potential (E_{pa}) of the POAP film could be oxidized easily. Hydroquinone oxidizes reversibly at the POAP-modified electrode showing a negative catalytic effect, which was attributed to the film resistance.

The electrochemistry of POAP-modified electrodes in the presence of the HQ/Q redox couple was extensively studied by RDEV in [45]. The effects of the electrode rotation rate (Ω), film thickness (ϕ), charge and concentration (c) of the electroactive species and acid concentration (x) of the redox active solution on the steady-state current-potential curves (*I-E*) were analyzed there in detail. Experimental results relating to the diffusion of these electroactive species through POAP were interpreted on the basis of the membrane-diffusion theory [46]. The electron-hopping model was invoked in order to obtain a diffusion constant for the electron transport inside the polymer film [25]. *I-E* curves, at different Ω values, for a 60 nm thick POAP film contacting an x M $HClO_4$ + (2-x) M $NaClO_4$ + c (HQ/Q) solution, where $x = 0.1$ and $c = 3 \times 10^{-3}$ M, are shown in [45] (Figure 1). Diffusion-limited currents ($I_{lim,a}$) were observed at $E > 0.8$ V (*versus* SCE) for HQ oxidation and at $E < 0.0$ V (*versus* SCE) for Q reduction ($I_{lim,c}$). Besides, main anodic and cathodic plateaux, the presence of a pre-plateau (0.2 V < E < 0.4 V) during Q reduction was also observed. At Ω and c fixed, *I-E* curves are film thickness and acid concentration dependent (Figure 2 and Figure 3, respectively). HQ oxidation is not thermodynamically mediated on POAP. Then, the observed anodic currents were assigned to the oxidation of HQ species that penetrate through the POAP films to reach the metal surface. Benzoquinone (Q) is reduced in two waves.

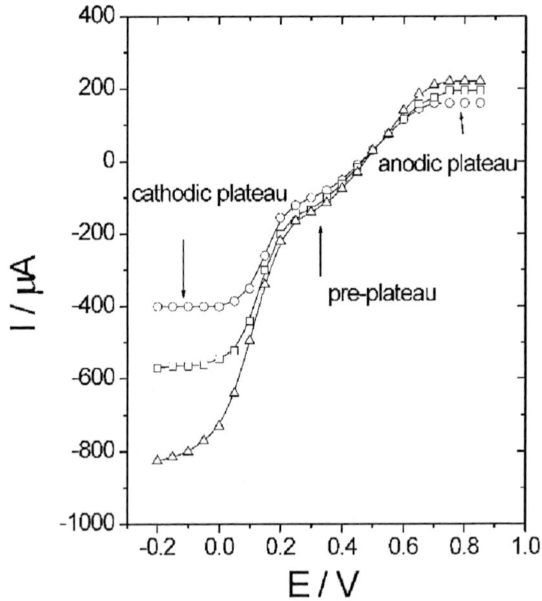

Figure 1. Steady-state current-potential (*I-E*) curves for different rotation rates of the rotating disc electrode: Ω= 500 (O), 1500 (□) and 3650 (Δ) rpm. Film thickness ϕ= 60 nm. Electrolyte solution 0.1 M $HClO_4$ + 1.9 M $NaClO_4$ + 3 x 10^{-3} M (HQ/Q).

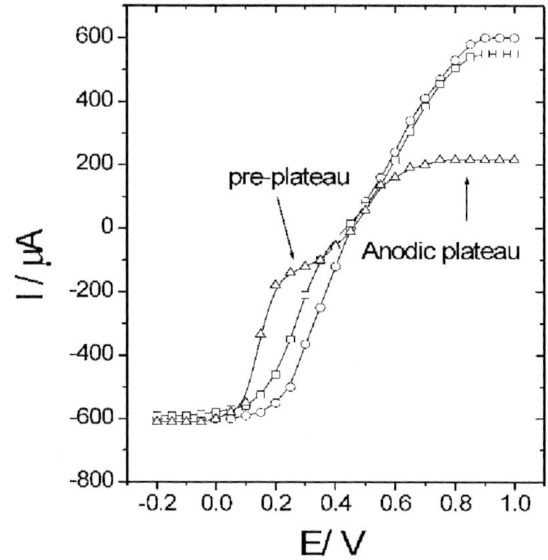

Figure 2. *I-E* curves for different film thickness: $\phi = 0$ (O), 13.7 (□) and 60 (Δ) nm. $\Omega = 1800$ rpm. Electrolyte solution 0.1 M $HClO_4$ + 1.9 M $NaClO_4$ + 3 x 10^{-3} M (HQ/Q).

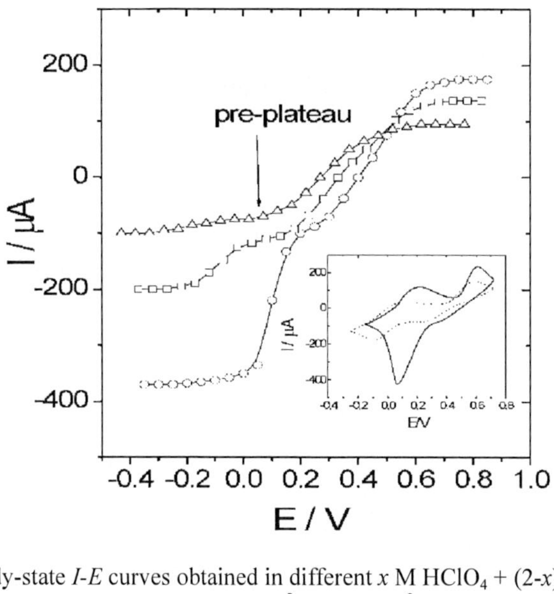

Figure 3. Steady-state *I-E* curves obtained in different *x* M $HClO_4$ + (2-*x*) M $NaClO_4$ + 3 x 10^{-3} M (HQ/Q) solutions, $x = 0.1$ (O), 10^{-2} (□) and 10^{-3} (Δ). $\Omega = 800$ rpm. $\phi = 60$ nm. Inset: Cyclic voltammograms for $x = 0.1$ (—) and 0.01(- - -), $v = 0.1$ Vs^{-1}.

The first wave (0.2 V < E < 0.4 V) originates from the penetration of reactants into the film coupled with their subsequent discharge at the metal|polymer interface. The permeation characteristics of POAP films during HQ oxidation, and Q reduction processes within the potential range corresponding to the pre-plateau, were analyzed in [45] by means of the Koutecky-Levich equation (Eq. (1) [46]):

$$I_{\lim}^{-1} = \phi\,(nFA\,\kappa D_{pol}\,c)^{-1} + (0.62\,nFAD^{2/3}\,\nu^{-1/6}\,\Omega^{1/2}\,c)^{-1} \quad (1)$$

In order to apply Eq. (1), *I-E* curves for different film thickness between 10 and 70 nm were recorded varying both Ω and the concentration c of the redox active species. Eq.(1) assumes that the electroactive species dissolves into the polymer film with a partition equilibrium $\kappa = c_{pol}/c$ at the film|solution interface, where c_{pol} and c are the respective concentrations of the species in the polymer and solution. D_{pol} and D are the diffusion coefficients of the electroactive species in the polymer and solution, respectively, $I_{\lim}$ is the limiting current for the oxidation or reduction of electroactive species and the other terms have their usual meanings.

At constant c, x and ionic strength of the solution, experimental $I_{\lim,a}^{-1}$ *vs.* $\Omega^{1/2}$ (Figure 4) and I_p^{-1} *vs.* $\Omega^{1/2}$ (Figure 5) representations for different film thickness higher than 13 nm were obtained in [45].

These representations show linear diagrams with nearly the same slope but different intercepts on the ordinate axis. Eq.(1) was employed in [45] to extract the parameter κD_s, which is a measure of the permeation rate of the HQ and Q species into POAP films. Authors of [45] determined the intercept at $\Omega \rightarrow \infty$ of the Koutecky-Levich (K-L) plots at each film thickness.

Then, κD_s values for both the HQ oxidation and the Q reduction processes were extracted from the first term of Eq. (1). As can be seen from Table 1 and Table 2, for both HQ and Q species, the κD_s value decreases with increasing film thickness. However, it was observed that for film thickness higher than 60 nm, κD_s becomes a constant independent of ϕ. Thus, extrapolated κD_s values, at high film thickness, were considered in [45] as true permeation rates for HQ and Q species across a POAP film.

The resultant extrapolated values are: κD_s (HQ) = (0.92 ±0.2) x 10^{-8} cm^2 s^{-1} and κD_s (Q) = (0.42 ±0.2) x 10^{-8} cm^2 s^{-1}. These κD_s values obtained in [45] for the permeation of HQ and Q species across POAP films are at least three orders of magnitude lower than the diffusion coefficient values of the same species in solution (D (HQ) = 1.5 x 10^{-5} cm^2 s^{-1}).

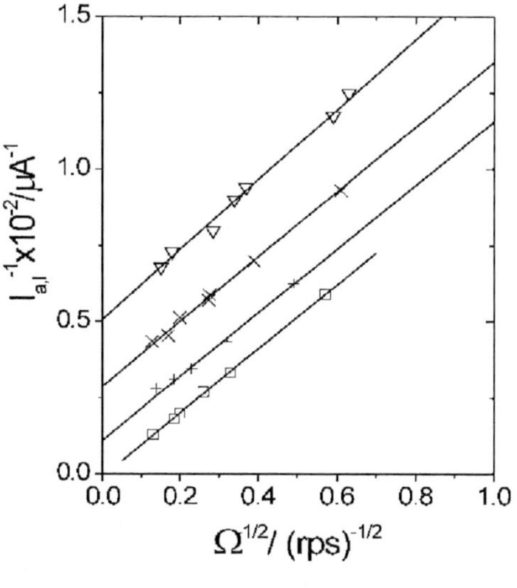

Figure 4. Koutecky-Levich plots ($I_{lim,a}^{-1}$ *vs.* $\Omega^{-1/2}$) for different film thicknesses: $\phi = 8$ (□), 41 (+), 60 (X) and 69 (∇) nm. Electrolyte solution 0.1 M $HClO_4$ + 1.9 M $NaClO_4$ + 3 x 10^{-3} M (HQ/Q).

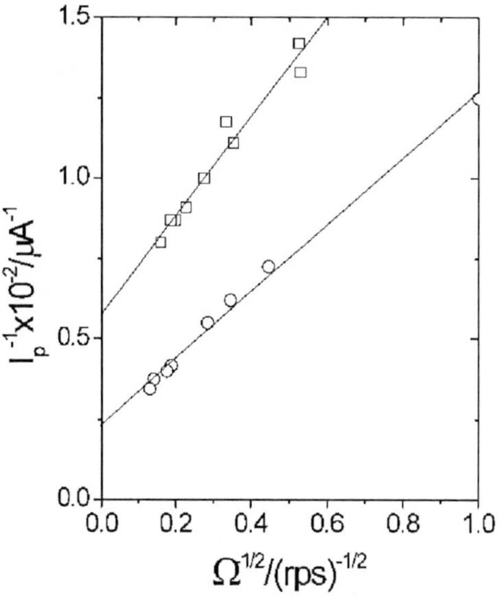

Figure 5. Reciprocal Levich plots (I_p^{-1} *vs.* $\Omega^{1/2}$) for the pre-plateau: $\phi = 41$ nm (o) and 60 nm (□). Electrolyte solution 0.1 M $HClO_4$ + 1.9 M $NaClO_4$ + 3 x 10^{-3} M (HQ/Q).

Table 1. HQ diffusion coefficient (*κDs*) values extracted from the first term of Eq. (1)

Polymer thickness, ϕ_p/nm	$10^8(\kappa D_s \pm 0.2, cm^2 s^{-1}$
13	6.27
22	4.50
41.7	2.60
60	1.52
69	1.04

HQ oxidation (E > 0.8 V *vs.* ECS), 0.1 M $HClO_4$ + 1.9 M $NaClO_4$ + 3 x 10^{-3} M HQ/Q solution, Electrode area: 0.23 cm^2, number of transferred electrons, $n = 2$.

Table 2. *κDs* values extracted from the first term of Eq. (1)

Polymer thickness, ϕ_p/nm	$10^8(\kappa D_s \pm 0.2, cm^2 s^{-1}$
41.7	1.25
60	0.75
69	0.61

Q reduction (0.2 < E < 0.4 V *vs.* ECS), 0.1 M $HClO_4$ + 1.9 M $NaClO_4$ + 3 x 10^{-3} M HQ/Q solution, Electrode area: 0.23 cm^2, number of transferred electrons, $n = 2$.

The POAP film thickness effect on the permeation rate of electroactive species was attributed in [45] to different polymer structures exhibited during film growth. In this regard, more dense structures as ϕ increases restrain the physical diffusion process of HQ and Q species through the film. Also, it was demonstrated in [45] that external electrolyte contacting the polymer film can be incorporated into the polymer phase, thus influencing the dynamics of electrochemical processes occurring within the film and at the film|electrode interface. In this regard, experiments were carried out with two film thickness (37 and 60 nm) contacting solutions, where the concentration of the redox couple (HQ/Q) was changed between 0.2 x 10^{-3} M and 12 x 10^{-3} M. In agreement with Eq. (1), at constant ϕ both the slope and the intercept at $\Omega \rightarrow \infty$ of experimental I_{lim}^{-1} *vs.* $\Omega^{1/2}$ representation decrease with increasing c. Then again, from the intercept on the ordinate axis for the different c values, κD_S as a function of c for both film thicknesses were obtained. While for the 60 nm thick film κD_S for the HQ oxidation remains (0.94 ± 0.2) x 10^{-8} cm^2 s^{-1} with increasing c, this value for the 37 nm thick film increases systematically with c (Table 3).

Table 3. κDs values extracted from the first term of Eq. (1)

10^3 c /mol L^{-1}	$10^8(\kappa D_s \pm 0.2, cm^2\ s^{-1}$
0.2	1.51
1.5	2.10
3	2.87
8	3.98
12	4.37

HQ oxidation (E > 0.8 V *vs.* ECS), 0.1 M $HClO_4$ + 1.9 M $NaClO_4$ + c M HQ/Q solution, Electrode area: 0.23 cm^2, number of transferred electrons, n = 2, film thickness, 37 nm.

Then, authors of [45] conclude that while thin films incorporate external electrolyte, the κD_s value obtained at high film thickness represents a true permeation rate dependent on the nature of the species incorporated into the polymer film but independent of both its concentration in the external electrolyte phase and the film thickness.

It was also observed in [45] that while K-L plots for HQ oxidation (Figure 4) show the same slope for all film thicknesses, in the potential range corresponding to the pre-plateau, for high film thickness, K-L plots exhibit a slightly higher slope (minor current) than those corresponding to thin ones (Figure 5). This effect was assigned to the absorption of Q into thick POAP films, which produces blocking effects. In this regard, authors of [45] report that there is evidence that in aqueous solutions while HQ is oxidatively chemisorbed on electrode surfaces, the interaction of Q is a simple chemisorption, that is, neither oxidative nor reductive [47].

Although authors of [45] have no evidential basis for separating the κD_s product, they compare this parameter with diffusion coefficient values of anions (D_x) required to maintain electroneutrality during the redox process of POAP. Considering the $D_x(ClO_4^-)$ value 2.32 x 10^{-7} cm^2 s^{-1} reported for perchlorate anions in a *pH* 1 and 2 M ionic strength solution [48], the sequence κD_s (HQ) < D_x (ClO_4^-) results. The perchlorate anion moves faster than HQ species. However, authors of [45] indicate that the difference could also be attributed to a partition coefficient, κ different from the unity for the active redox species. The permeability of HQ was also compared with that of other charged species such as $Fe(CN)_6^{-4}$. It was observed that the permeability of HQ species at POAP films is higher than that of $Fe(CN)_6^{-4}$ species. This behaviour was explained in terms of different types of interaction of these species with the POAP film. Both HQ and Q species are able to form hydrogen bonds with the polymer, which in turn would lead to their penetration into the bulk polymer. However, a stronger interaction due to ion pairing effects is expected in the presence of $Fe(CN)_6^{-4}$ species, as compared with hydrogen bonds formed during HQ-POAP interaction. These different interactions lead to a lower mobility of $Fe(CN)_6^{-4}$ anion. The higher κD_s value for HQ as compared with that of Q was explained in terms of the different degree of protonation of the reduced and oxidized states of POAP. POAP is slightly more protonated in the reduced state than in the oxidized one [24]. This is due to the fact that the imino groups formed during oxidation are not protonated. Then Q more easily forms hydrogen bonds between protonate nitrogen and oxygen than HQ in the oxidized state of POAP. This, in turn, would reduce the diffusion rate of Q as compared with HQ.

The limiting current for Q reduction at $E < 0.0$ V is independent of ϕ (Figure 2) and follows the Levich equation (Eq.(2)). This wave was associated with an electron-transfer mediation at the POAP|HQ/Q interface [49], where Q reduces to HQ.

$$I_{\text{lim}} = 0.62\, nFAc\, D^{2/3}\, v^{-1/6}\, \Omega^{1/2} \tag{2}$$

In Eq. (2), I_{lim} is the mass-transfer limited current, n is the number of electrons transferred during the redox mediation, A the electrode area, F the Faraday constant, D the diffusion coefficient of the electroactive species (Q) in solution, v the kinematic viscosity of the solution, c is the bulk concentration of the electroactive species in solution and Ω is the electrode rotation rate.

2.2. Deactivated POAP Films in Contact with a Solution Containing the HQ/Q Redox Couple

In another work [50] POAP films were deactivated to study how their permeation and electron-transport processes are modified by the degree of deactivation. To this end different POAP films were immersed in a 50 mM $Fe_2(SO_4)_3$ solution for a different time period in order to obtain different degrees of deactivation (Table 4). Deactivation of POAP films by contact with a ferric cation solution was attributed to the interaction of iron ions with the redox sites of POAP, which impedes the protonation reaction of the polymer [43].The degree of deactivation of each POAP film employed in [50] was assessed by comparing its voltammetric response after being subjected to a treatment with the ferric cation solution with the response of the non-deactivated film. Figure 6 shows voltammetric responses corresponding to the films (1), (4) and (7) indicated in Table 4, after being subjected to the soaking process previously described.

Table 4. Voltammetric reduction charge $Q_{Red,c}$ and degree of deactivation, θ_c, of different POAP films

POAP films [a]	Soaking time [b]/h	$Q_{Red,c}$ [c]/ mCcm^{-2}	θ_c [d]
1	2	2.44	0.13
2	4	2.04	0.18
3	5	1.87	0.33
4	10	1.57	0.44
5	16	1.32	0.51
6	21	1.15	0.56
7	24	1.12	0.60
8	33	1.06	0.62

[a] Numbers 1 to 8 represent different deactivated POAP films.

[b] Different soaking times in a 0.1 M H_2SO_4 + 0.05 M $Fe_2(SO_4)_3$ solution.

[c] Voltammetric reduction charge of the different deactivated POAP films after being subjected to the soaking process in the presence of the ferric cation solution.

[d] Degree of deactivation of the different POAP films calculated from $\theta_c = 1-(Q_{Red,c}/Q_{Red,T})$. $Q_{Red,T}$ (= 2.8 mC cm^{-2}) is the voltammetric reduction charge of a non-deactivated film.

The more attenuated voltammetric response of a POAP film with increasing soaking time in the ferric cation solution was considered as representative of a higher degree of deactivation of the polymer film. In this regard, voltammetric reduction charge values corresponding to the completely reduced POAP films were compared for both non-deactivated films ($Q_{Red,T}$ = 2.8 mC cm^{-2}) and films subjected to the soaking process in the presence of the Fe(III) solution ($Q_{Red,c}$) (see column 3 in Table 4); then, a degree of deactivation (column 4 of Table 4) was defined as

$$\theta_c = 1 - (Q_{Red,c}/Q_{Red,T}) \quad (3)$$

where $Q_{Red,c}$ is the total reduction charge assessed by integration of the corresponding j-E response from E = 0.5 V towards the negative potential direction for a deactivated film, and $Q_{Red,T}$ = 2.8 mC cm^{-2} is the total reduction charge for the non-deactivated film. In this way, for a non-deactivated POAP film (dashed curve in Figure 6) the degree of deactivation was $\theta_c = 0$, taking $Q_{Red,T} = 2.8$ mC cm^{-2} as reference charge. However, values of $\theta_c > 0$ are indicative of POAP films that have been deactivated. The deactivation process of POAP by contact with a ferric cation solution was chosen in [50] because it has been previously demonstrated by CV [43] that the initially deactivated polymer can be partially reactivated by treatment with a 0.1 M NH_4OH solution. Then, RDEV experiments were carried out in [50] with deactivated POAP films in contact with a solution containing the HQ/Q redox couple. By analyzing variations in the cathodic and anodic limiting currents with the electrode rotation rate, charge-transport parameters as a function of the degree of deactivation were extracted.

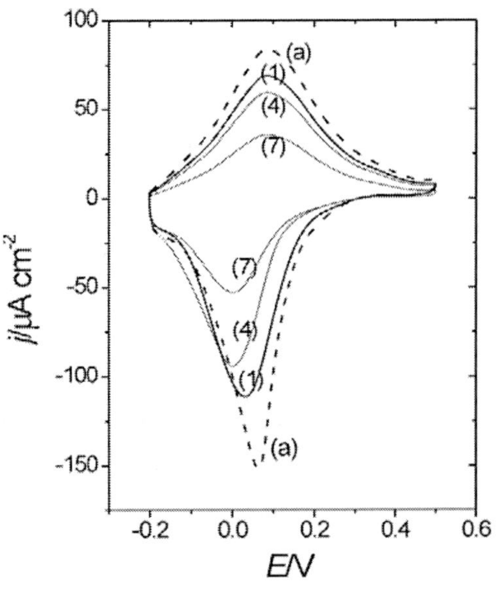

Figure 6. Voltammetric (j-E) responses for 2.8 mC cm^{-2} thick POAP films: (a) (- - - -) a freshly prepared POAP film. Numbers (1), (4) and (7) correspond to POAP films with different degrees of deactivation (θ_c): (1): $\theta_c = 0.13$; (4): $\theta_c = 0.44$; (7): $\theta_c = 0.60$. Electrolyte: 0.1 M $HClO_4$ + 0.4 M $NaClO_4$. Scan rate: $v = 0.01$ V s^{-1}.

Figure 7 shows I-E curves at different electrode rotation rates for a POAP film with $\theta_c = 0.13$. The anodic limiting current for a deactivated POAP film is lower than that for a non-deactivated one at the same electrode rotation rate and redox couple concentration. Also, after a given electrode rotation rate, which depends on the degree of deactivation, the cathodic limiting current for a deactivated film becomes independent of this variable (Figure 7). Then, at potential values $E < 0.0$ V, $I_{\mathrm{lim,c}}$ *versus* $\Omega^{1/2}$ dependences, for a non-deactivated film and also for each one of the deactivated POAP films indicated in Table 4, were compared in [50]. For a non-deactivated POAP film a linear $I_{\mathrm{lim,c}}$ *versus* $\Omega^{1/2}$ dependence, which follows the Levich equation, is obtained within a wide range of Ω values (Figure 8). However, for POAP films that have been deactivated, after a certain Ω value, a constant cathodic limiting current value $I_{\mathrm{lim,c}}$ independent of Ω is achieved. From Figure 8, it is observed that the transition at which the cathodic limiting current becomes independent of Ω occurs at lower Ω values as the degree of deactivation increases. Interpretation of this effect was made on the basis of the electron-hopping model [25,49,51] by employing Eq. (4)

$$I_e = n\,F\,A\,D_e\,c^*/\phi \qquad (4)$$

According to Eq. (4), the limiting current value at which $I_{\mathrm{lim,c}}$ (= I_e) becomes independent of Ω is related to the maximum flux of redox species confined in the polymer. Eq. (4) implies a constant gradient of redox species across the film (c^*/ϕ), where c^* is the concentration of redox sites of the polymer and ϕ the film thickness. D_e represents a measure of the electron hopping rate and n express the numbers (fractions) of unit charges per monomer unit of the polymer. A is the electrode area and F the Faraday's constant. Experimental I_e values, corresponding to different deactivated POAP films contacting a 2×10^{-3} M HQ/Q solution, were extracted in [50] from the cathodic plateau and they were represented as a function of θ_c (Figure 9). I_e decreases with increasing θ_c. In this regard, such a constant value of the current (I_e) at a given Ω value for deactivated POAP films was related to a slow electron transport across the POAP film to mediate in the electron-transfer reaction at the POAP|HQ/Q interface, as compared with a non-deactivated POAP film. This effect was explained in terms of the increase in the flux of Q with the increase in Ω.

If the flux of Q exceeds the supply of electrons from the electrode through the polymer to the electrolyte interface, the rate-limiting step will shift from the limiting transport of Q to the limiting transport of the charge through the polymer. In order to verify this limiting charge-transport process across the POAP film, the HQ/Q concentration in solution was also varied in [50].

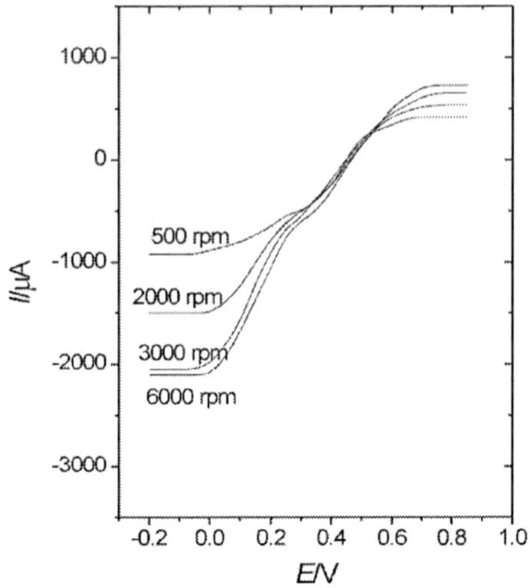

Figure 7. Steady-state current-potential (*I-E*) curves for different rotation rates (Ω) of the rotating disc electrode. Deactivated POAP film ($\theta_c = 0.13$). Ω Values are indicated in the figure. Film thickness: 60 nm. Electrolyte: 0.1 M $HClO_4$ + 0.4 M $NaClO_4$+ 2 x 10^{-3} M(HQ/Q).

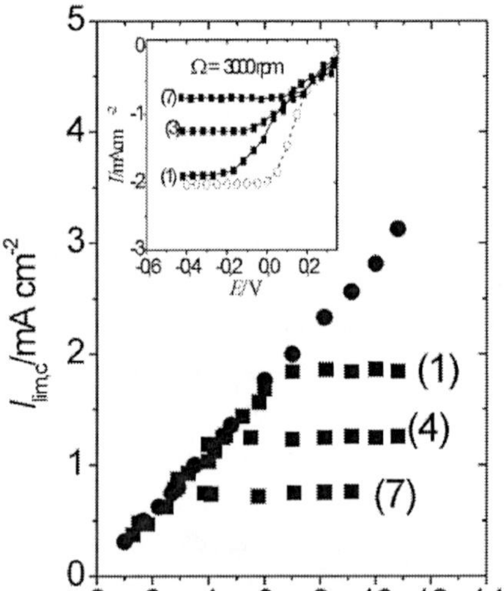

Figure 8. Levich representations $I_{lim,c}$ *vs.* $\Omega^{1/2}$ for POAP films contacting a 0.1 M $HClO_4$ + 0.4 M $NaClO_4$+ 2 x 10^{-3} M (HQ/Q) solution. (●) A non-deactivated POAP film. (1), (4) and (7) correspond to the three films indicated in Figure 6. Inset: steady-state (*I-E*) curves for potential values $E < 0.2$ V at $\Omega = 3000$ rpm.

Figure 10 shows the $I_{lim,c}$ *versus* $\Omega^{1/2}$ dependence for the same films indicated in Figure 8, but contacting a 3 x 10^{-3} M HQ/Q solution. The inset in Figure 10 shows the same dependence for a 4 x 10^{-3} M HQ/Q solution. By comparing Figure 8 and Figure 10, the constant current for a given deactivated film remains unchanged with increasing redox couple concentration in solution. However, this constant current is reached at a lower electrode rotation rate as the concentration of HQ/Q in solution is increased.

According to Eq. (4), the slower electron transport in deactivated films, as compared with non-deactivated ones, was attributed to a decrease of D_e. From data shown in Figure 9 for I_e, and using the corresponding parameter values for POAP (c = 4.7 M [18,19,21], A = 1 cm^2, n = 0.44 [23] and ϕ_p = 60 nm) in Eq. (4), a relative decrease of D_e from 5.4 x 10^{-11} to 2.4 x 10^{-11} cm^2 s^{-1} for a relative increase of θ_c from 0.13 to 0.62 was obtained in [50]. The decrease of D_e was attributed to an increase of the hopping distance between remnant active redox sites after deactivation [52].

The increase of the hopping distance with deactivation is consistent with the electronic interaction of iron ions with redox sites of POAP proposed in [43], which hinder the acceptance of protons by some redox sites, and then they become inactive for the redox reaction of the polymer.

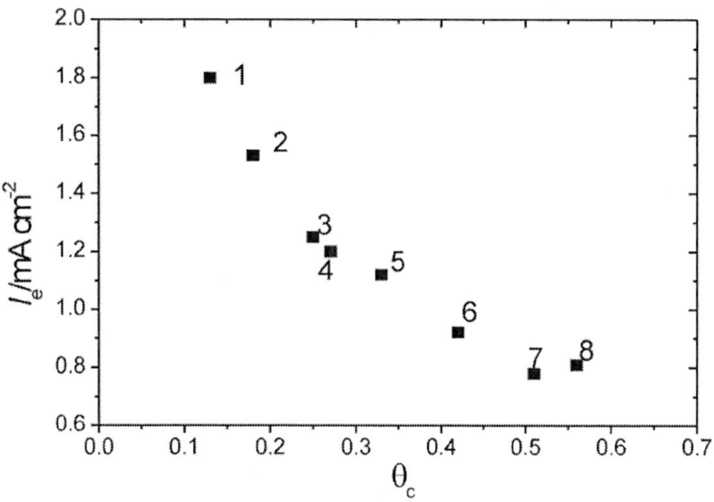

Figure 9. Electron current I_e (Eq. (4)) as a function of θ_c. Numbers 1 to 8 correspond to each one of the POAP films indicated in Table 4.

With regard to the anodic plateau, Eq. (1) was also employed in [50] to study the effect of the degree of deactivation of POAP on the diffusion rate of

HQ species. In the same way as with non-deactivated film, at constant c, experimental $I_{a,l}^{-1}$ *versus* $\Omega^{-1/2}$ representations for deactivated POAP films give linear diagrams with the same slope but different intercepts on the ordinate axis (Figure 11). Then κD_s values for deactivated films were obtained in [50] from the first term of Eq. (1).

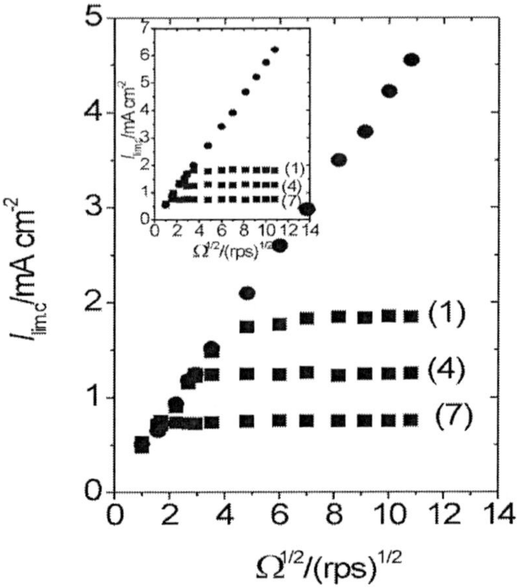

Figure 10. Levich representations $I_{lim,c}$ *vs*. $\Omega^{½}$ for POAP films contacting a 0.1 M $HClO_4$ + 0.4 M $NaClO_4$ + 3 x 10^{-3} M (HQ/Q) solution. (●) A non-deactivated POAP film. (1), (4) and (7) correspond to the three films indicated in Figure 6. Inset: $I_{lim,c}$ *vs*. $\Omega^{½}$ representations for the same films indicated in Figure 6, but contacting a 0.1 M $HClO_4$ + 0.4 M $NaClO_4$.+ 4 x 10^{-3} M (HQ/Q) solution.

As can be seen from Table 5, κD_s decreases as the degree of deactivation increases. Again, the author of [50] cannot separate κ from D_s. However, he considers that for a deactivated film some redox sites are inactive and cannot accept protons, then it is possible that the high degree of swelling, as a direct result of solvent ingress to compensate for changes in the charge density within the film (proton inclusion and/or anion exclusion during reduction and the reverse during oxidation), will be absent in a deactivated film. Then, as θ_c increases the poor electrolyte permeation would lead to low κ values, which would make the solute electroactive diffusion within the film difficult.

Table 5. Diffusion rate, κD_s, of hydroquinone species at different degrees of deactivation, θ_c

θ_c	10^8 κD_s [a] /cm^2 s^{-1}
0	2.0
0.13	1.24
0.42	0.88

[a] Diffusion rate values of hydroquinone species extracted from Koutecky-Levich plots.

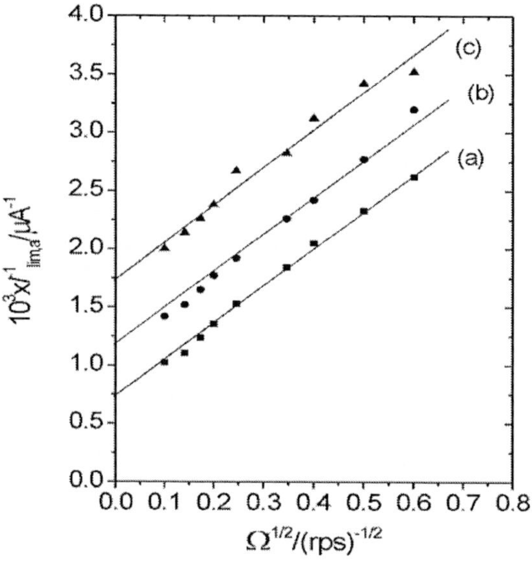

Figure 11. Koutecky-Levich plots ($I_{a,l}^{-1}$ *vs*. $\Omega^{1/2}$) for different deactivated POAP films. (a) $\theta_c = 0$; (b) $\theta_c = 0.13$; (c) $\theta_c = 0.27$. Film thickness: 60 nm. Electrolyte: 0.1 M $HClO_4$ + 0.4 M $NaClO_4$ + 2 x 10^{-3} M (HQ/Q) solution. $E = 0.8$ V.

Another experiment was performed in [50] with POAP films that, after being deactivated by treatment with the ferric cation solution, were treated with an ammonium solution. Figure 12 compares the voltammetric response of a non-deactivated film with those of the same film, firstly deactivated by soaking in a ferric cation solution and then reactivated by treatment with a 0.1 M NH_4OH solution. An enhancement of the redox response is observed after reactivation. Figure 13 compares the $I_{lim,c}$ *versus* $\Omega^{1/2}$ dependence for the non-deactivated film (plot (a)) with those of the same film firstly deactivated (plot (b)) and then, reactivated (plot (c)) by alkaline treatment.

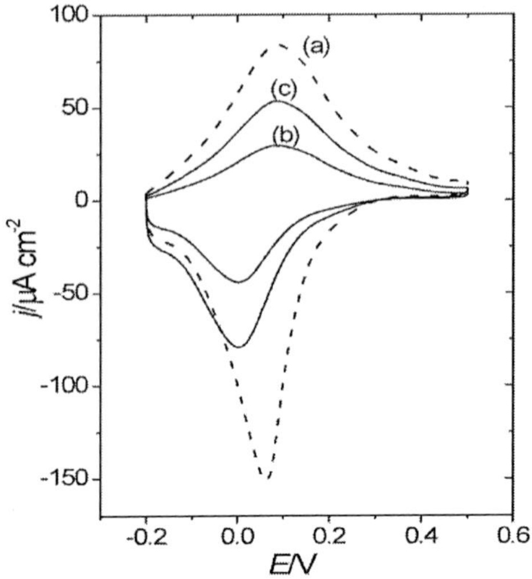

Figure 12. Voltammetric (*j*-*E*) responses of: (a) A non-deactivated POAP film, $\theta_c = 0$; (b) the same film deactivated, $\theta_c = 0.62$; (c) the same film after reactivation, $\theta_c = 0.43$. Electrolyte: 0.1 M $HClO_4$ + 0.4 M $NaClO_4$. Scan rate: $\nu = 0.01$ V s^{-1}.

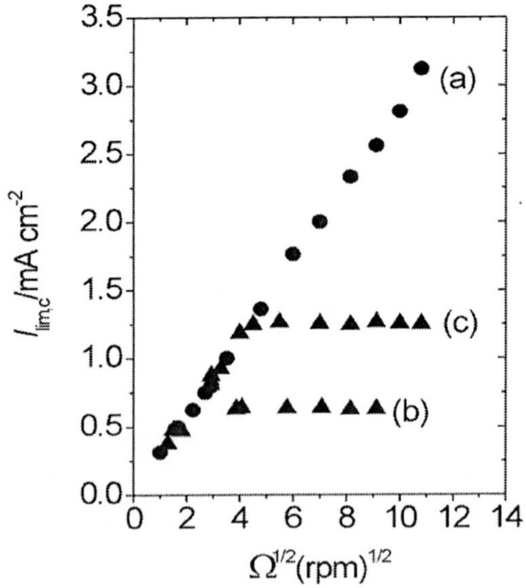

Figure 13. Levich representations $I_{\text{lim,c}}$ *vs*. $\Omega^{1/2}$ for the same POAP film indicated in Figure 12 but contacting a 0.1 M $HClO_4$ + 0.4 M $NaClO_4$+ 2 x 10^{-3} M (HQ/Q) solution. E = - 0.2 V. (a) $\theta_c = 0$; (b) $\theta_c = 0.62$; (c) $\theta_c = 0.43$.

As can be seen from Figure 13 also, an enhancement of the constant cathodic limiting current value is obtained with reactivation, but the Levich dependence corresponding to a non-deactivated film is not totally recovered.

Figure 14 shows the $I_{lim,c}$ *versus* $\Omega^{1/2}$ dependence for the same film indicated in Figure 13, firstly deactivated and then reactivated, but for two different HQ/Q concentrations.

Again, it is observed that the constant current values, I_e, remain unchanged for a given degree of deactivation, but they are reached at a lower electrode rotation rate as the redox couple concentration in solution is increased. Then, the electron conduction of a deactivated POAP film can only be partially recovered with reactivation by treatment with an alkaline solution.

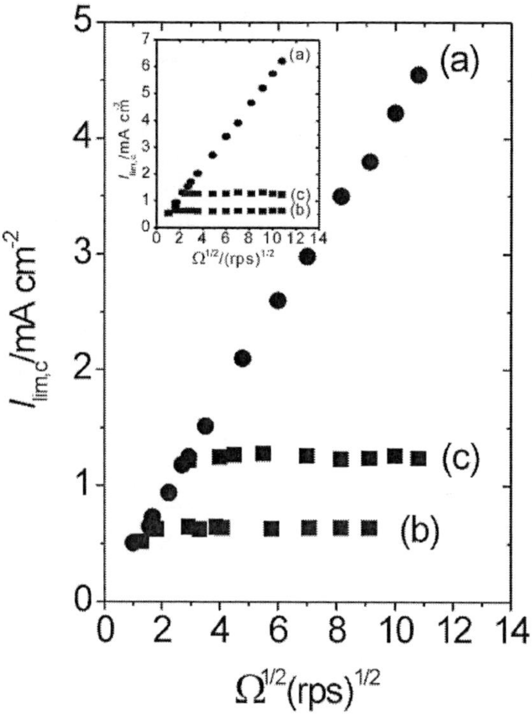

Figure 14. Levich representations $I_{lim,c}$ *vs.* $\Omega^{½}$ for the same POAP film indicated in Figure 12 but contacting a 0.1 M $HClO_4$ + 0.4 M $NaClO_4$+ 3 x 10^{-3} M (HQ/Q) solution. E = - 0.2 V. (a) $\theta_c = 0$; (b) $\theta_c = 0.62$; (c) $\theta_c = 0.43$. Inset: Levich representations $I_{lim,c}$ *vs.* $\Omega^{½}$ for the same POAP film indicated in Figure 12 but contacting a 0.1 M $HClO_4$ + 0.4 M $NaClO_4$+ 4 x 10^{-3} M (HQ/Q) solution. The same θ_c values as the above-mentioned.

3. The Electrochemical Behaviour of Deactivated POAP Films in the Presence of the HQ/Q Redox Couple Studied by Electrochemical Impedance Spectroscopy (EIS)

The electrochemical behaviour of deactivated POAP films in the presence of the HQ/Q redox couple was studied by EIS in [53]. POAP films were deactivated by employing the same procedure described in [50]. The objective of this work was to obtain a more complete series of transport parameters of deactivated POAP films than that obtained previously from RDEV in [50], that is, interfacial metal-film and film-solution resistances ($R_{m|f}$, $R_e^{f|s}$, $R_i^{f|s}$), the high-frequency capacitance (C_H), the redox capacitance (C_p) and the electron and ion diffusion coefficients (D_e, D_i). The same series of deactivated electrodes (see Table 4) employed in [50] were used in [53]. Dependences of the transport parameters on the degree of deactivation were extracted in [53] from the fitting of impedance diagrams by employing the Vorotyntsev's model described in [15]. Deactivated POAP films were also reactivated by treatment with a 0.1 M NH_4OH solution and then, *ac* impedance measurements were carried out in [53] to obtain the corresponding values of the transport parameters. The theory of Vorotyntsev et al. [15] was developed within the framework of the assumption that the redox active species are only present in the solution phase but not inside the film, and they participate in the interfacial electron exchange with the polymer at the film-solution boundary. Then, only impedance diagrams at negative potential values ($E < 0.0$ V), where POAP acts as mediator in the Q reduction, were analyzed in [53].

3.1. Impedance Measurements at the Cathodic Plateau (E < 0.0 V)

Impedance measurements with non-deactivated POAP films and POAP films with different θ_c values were compared in [53]. The measurements were performed, firstly, only in the presence of the supporting electrolyte solution (0.1 M $HClO_4$ + 0.4 M $NaClO_4$), and then, in a solution of equimolar concentrations of *p*-benzoquinone and hydroquinone species (0.1 M $HClO_4$ + 0.4 M $NaClO_4$ + 2 x 10^{-3} M (Q/HQ)). Nyquist diagrams of a non-deactivated POAP film in contact with the supporting electrolyte exhibit a Warburg region at high frequency followed by a pseudo-capacitive increase [54] (Figure 15). However, a high-frequency semicircle is observed in the impedance diagram of a deactivated POAP film (Figure 16).

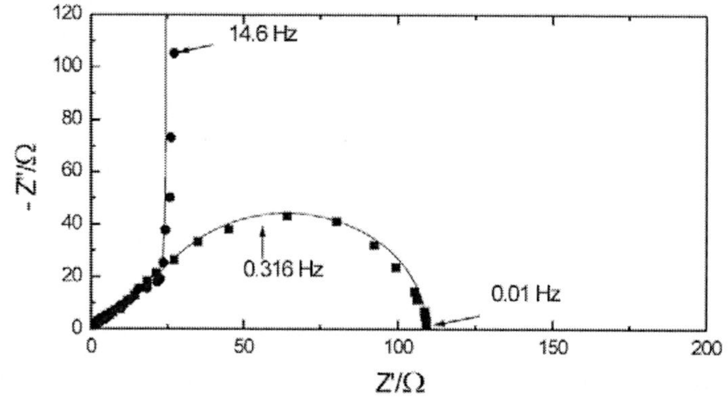

Figure 15. *ac* Impedance diagrams in the Nyquist coordinates (-*Z*'' vs. *Z*) obtained at *E* = -0.3 V for a non-deactivated POAP film (●) contacting a 0.1 M $HClO_4$ + 0.4 M $NaClO_4$ solution. The same POAP film (■) contacting a 0.1 M $HClO_4$ + 0.4 M $NaClO_4$ + 2 x 10^{-3} M (HQ/Q) solution. Rotation rate, Ω= 300 rpm. Discrete points are experimental data and continuous lines represent the fitting by using the theory given in [15].

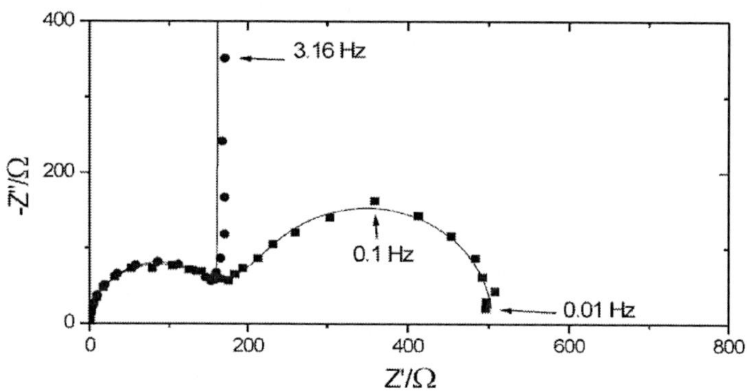

Figure16. *ac* Impedance diagrams in the Nyquist coordinates (-*Z*'' vs. *Z*) obtained at *E* = -0.3 V for a deactivated POAP film (●) contacting a 0.1 M $HClO_4$ + 0.4 M $NaClO_4$ solution. The same POAP film (■) contacting a 0.1 M $HClO_4$ + 0.4 M $NaClO_4$ + 2 x 10^{-3} M (HQ/Q) solution. Rotation rate, Ω= 600 rpm. θ_c = 0.33. Discrete points are experimental data and continuous lines represent the fitting by using the theory given in [15].

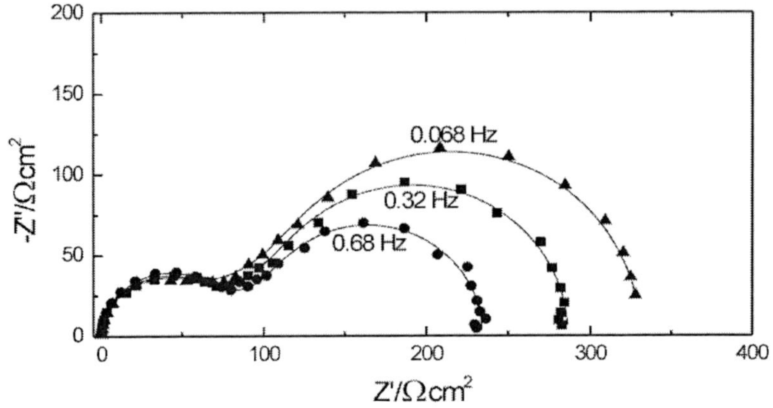

Figure 17. *ac* Impedance diagrams in the Nyquist coordinates (-Z'' vs. Z) obtained at *E* = -0.3 V for the film (1) of Table 4, $\theta_c = 0.13$. The different diagrams correspond to different electrode rotation rates, Ω: (▲) 1000 rpm; (■) 1500 rpm; (●) 2600 rpm. Electrolyte: 0.1 M $HClO_4$ + 0.4 M $NaClO_4$ + 2 x 10^{-3} M (HQ/Q) solution. Discrete points are experimental data and continuous lines represent the fitting by using the theory given in [15].

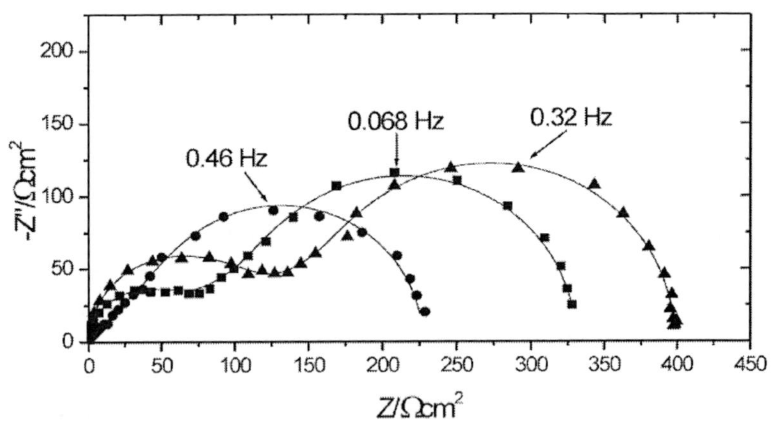

Figure 18. *ac* Impedance diagrams in the Nyquist coordinates (-Z'' vs. Z) obtained at *E* = -0.3 V and a constant electrode rotation rate, Ω= 1000 rpm, for: (●) a non-deactivated POAP film ($\theta_c = 0$); (▲) the same film after deactivation ($\theta_c = 0.56$); and (■) the same film after reactivation ($\theta_c = 0.31$). Electrolyte: 0.1 M $HClO_4$ + 0.4 M $NaClO_4$ + 2 x 10^{-3} M (HQ/Q) solution. Discrete points are experimental data and continuous lines represent the fitting by using the theory given in [15].

In the presence of the HQ/Q redox couple, while a non-deactivated POAP film only shows one loop at low frequency (Figure 15), a deactivated POAP film exhibits two loops (Figure 16). For deactivated POAP films, it was found that while the loop at low frequency is Ω dependent (Figure 17), the high-frequency semicircle is independent of this variable. However, the size of the high-frequency semicircle depends on the degree of deactivation (θ_c). In this regard, at a given Ω value, the higher the θ_c value is, the greater the high-frequency semicircle becomes (Figure 18).

3.2. Interpretation of AC Impedance Diagrams

The theory of Vorotyntsev leads to an impedance response described by Eq. (5)

$$Z_{m|film|es} = R_{m|f} + R_f + R_s + [Z_e^{f|s} R_i^{f|s} + W_f Z_{12}^{m}] (Z_e^{f|s} + R_i^{f|s} + 2\, W_f \coth 2\nu)^{-1} \quad (5)$$

where

$$Z_{12}^{m} = Z_e^{f|s} [\coth \nu + (t_e - t_i)^2 \tanh \nu] + R_i^{f|s}\, 4t_i^2 \tanh \nu + W_f\, 4t_i^2 \quad (6)$$

In Eq. (5) and Eq. (6)

$\nu = [(j\omega\phi_p^2)/4D]^{1/2}$ is a dimensionless function of the frequency ω, ϕ_p is the film thickness, D is the binary electron-ion diffusion coefficient, and t_i and t_e are the migration (high frequency) bulk-film transference numbers for anions and electrons, respectively. D is defined as $D = 2D_iD_e(D_e+D_i)^{-1}$ and $t_{i,e} = D_{i,e}(D_e+D_i)^{-1}$, where D_e and D_i are the diffusion coefficients for the electrons and ion species, respectively.

$W_f = [\nu/j\omega\phi_pC_p] = \Delta R_f/\nu$ is a Warburg impedance for the electron-ion transport inside the polymer film. ΔR_f ($=\phi_p/4DC_p$) is the amplitude of the Warburg impedance inside the film, and C_p is the redox capacitance per unit volume.

R_f ($=\phi_p/\kappa$) is the high-frequency bulk-film resistance, R_s the ohmic resistance of the bulk solution (κ is the high-frequency bulk conductivity of the film), $R_{m|f}$ is the metal|film interfacial electron-transfer resistance, and $R_i^{f|s}$ is the film|solution interfacial ion-transfer resistance.

$Z_e^{f|s} = (R_e^{f|s}+W_s)$ is the electronic impedance, where $R_e^{f|s}$ is the interfacial electron-transfer resistance at the film|solution interface and W_s is the convective diffusion impedance of redox species in solution, which contains the bulk concentrations of ox(red) forms, $c_{ox}(c_{red})$, and their diffusion coefficients inside the solution, $D_{ox}(D_{red})$. Also, it contains the Nernst layer thickness, δ.

$R_e^{f|s}$ is defined as

$$R_e^{f|s} = RT\,(nF^2 k_o c_{red})^{-1} \tag{7}$$

where k_o is the rate constant of the reaction between the film and the redox active forms in solution. Diffusion of the redox forms from the bulk solution to the film|solution interface can be regarded as stationary through the diffusion layer thickness, expressed in cm by

$$\delta = 4.98\, D_{ox,red}^{\;1/3}\, \eta^{1/6}\, \Omega^{-1/2} \tag{8}$$

where η is the kinematic viscosity of the solution in the same units as $D_{ox,red}$, and Ω the rotation rate of the disk electrode in rpm. The rest of the constants have their usual meaning. This model also includes the impedance behaviour of the polymer contacting the inactive electrolyte (absence of the redox couple in solution) by considering $Z_e^{f|s} \rightarrow \infty$ in Eq. (5).

3.3. Dependences of the Different Charge-Transport and Charge-Transfer Parameters on the Degree of Deactivation

Impedance spectra shown in [53] correspond to a potential value of -0.3 V. This potential value was considered as representative of the potential region where POAP is in its reduced state and acts as mediator in the reduction process of Q.

In all diagrams shown from Figure 15 to Figure 18, besides the measured values (discrete points), the simulated curves (continuous lines) calculated by using Eq. (5) are displayed. Real and imaginary parts *versus* frequency plots corresponding to each one of the Nyquist representations shown from Figure 15 to Figure 18 were also built in [53]. One of these representations for θ_c values of 0.33 is shown in Figure 19.

A good fitting was reported in [53] for all impedance diagrams. The fitting procedure by using Eq. (5) was based on the CNLS (Complex Nonlinear Squares) method. A rigorous fitting procedure was performed in [53] where six replicate measurements for each degree of deactivation were carried out. The error structure was assessed following the method recommended by Agarwal et al. [55-57] and Orazem [58]. The standard deviation for the real (σ_{Zr}) and imaginary (σ_{Zj}) parts of the impedance followed the form proposed by Orazem

$$\sigma_{Zr} = \sigma_{Zj} = \alpha\left|Zj\right| + \beta\left|Zr\right| + \gamma\left|Z\right|^2 Rm^{-1} + \delta \qquad (9)$$

where Rm is the current-measuring resistor used for the experiment, Zr is the real part of the impedance, and Zj is the imaginary part of the impedance. α, β, γ and δ are constants that have to be determined. The values of these scaling factors were $\alpha = 0$, $\beta = 4.89 \times 10^{-3} \pm 0.007 \times 10^{-3}$, $\gamma = 2.9 \times 10^{-5} \pm 0.1 \times 10^{-5}$ and $\delta = 7.2 \times 10^{-3} \pm 0.3 \times 10^{-3}$. The error structure was found to follow the same model within the range of degree of capture $0.05 < \theta_c < 0.65$. At θ_c values lower than 0.05, the error structure model parameters had different values, but these results are not reported here. Then, continuous lines in Figure 15 to Figure 19 represent the weighted complex non-linear least-squares fit to the data. The regression was weighted by the inverse of the variance of the stochastic part of the measurement. In all conditions the weighted sum of the square of the residuals was below one [58].

In the simulations performed in [53] the number of transferred electrons, n, was assumed to be 0.44 [23], and diffusion coefficient values of the redox species (Q and HQ) were considered equal, $D_{ox,red} = 1.5 \times 10^{-5}$ cm^2 s^{-1}. Also, the bulk concentrations of the redox substrate species were considered equal ($c_{ox} = c_{red} = 2 \times 10^{-6}$ mol cm^{-3}). The polymer thickness was considered as $\phi =$ 60 nm [45]. The value of the total redox site concentration of POAP was considered as $c_o = 4.7 \times 10^{-3}$ mol cm^{-3} [23]. The different parameters contained in Eq. (5) ($R_{m|f}$, $R_i^{f|s}$, $R_e^{f|s}$, C_p, D_e and D_i) were calculated in [53] from the experimental impedance data by the fitting procedure described above.

The first three parameters ($R_{m|f}$, $R_i^{f|s}$ and $R_e^{f|s}$) were varied without restraints during the fitting. However, some reference values were considered for C_p, D_e and D_i. For the freshly prepared POAP film thickness used in this work ($\phi_p = 60$ nm) and solution $pH = 1$, D_e and D_i values were allowed to vary within the range 10^{-7}-10^{-11} cm^2 s^{-1}, in such a way that diffusion coefficient values lower than 10^{-11} were considered not realistic for these thick films.

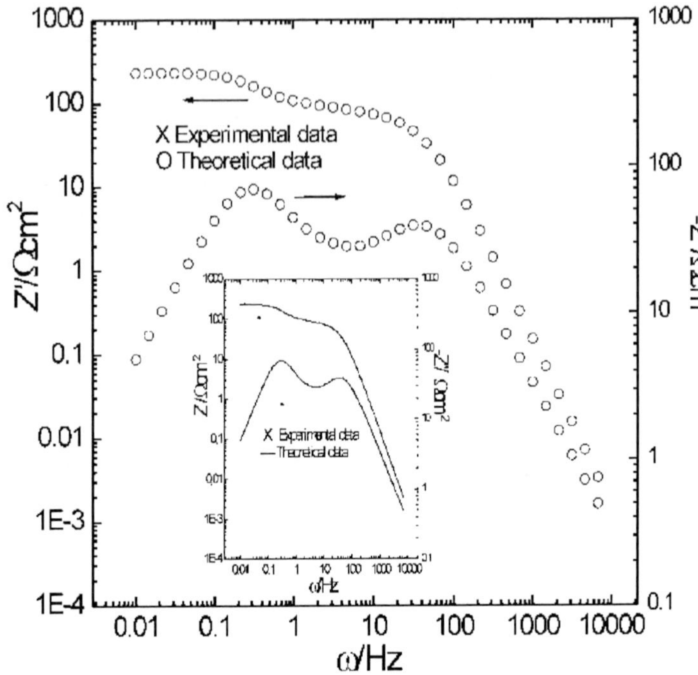

Figure 19. Real part (Z') and imaginary part ($-Z''$) versus frequency (ω) plots (point-to-point representations) for a POAP film with $\theta_c = 0.33$. $E = -0.3$ V, $\Omega = 5000$ rpm. Electrolyte: 0.1 M $HClO_4$ + 0.4 M $NaClO_4$ + 2 x 10^{-3} M (HQ/Q) solution. (X) Experimental data; (O) the fitting by using Eq. (5) [15]. Inset: the same representations indicated above but with theoretical values represented as continuous lines.

That is, D_e and D_i values lower than 10^{-11} were only obtained from impedance diagrams (not shown in [53]) of very thin POAP films ($Q_{T,Red} = 0.2$ mC cm^{-2}, $\phi_p = 10$ nm) contacting solutions of $pH = 1$, where incomplete coating of the metal area by the thin polymer film was considered as possible. Concerning C_p, reference values were extracted from experimental -Z" vs. ω^{-1} slopes of impedance diagrams at sufficiently low frequency (in the absence of the redox substrate in solution).

A contribution of the interfacial capacitance, C_H, also considered as a fitting parameter, was included in order to represent the actual impedance diagrams from the calculated ones.

The dependences of the different charge-transport and charge-transfer parameters on θ_c, extracted from the fitting procedure described above, are shown from Figure 20 to Figure 26.

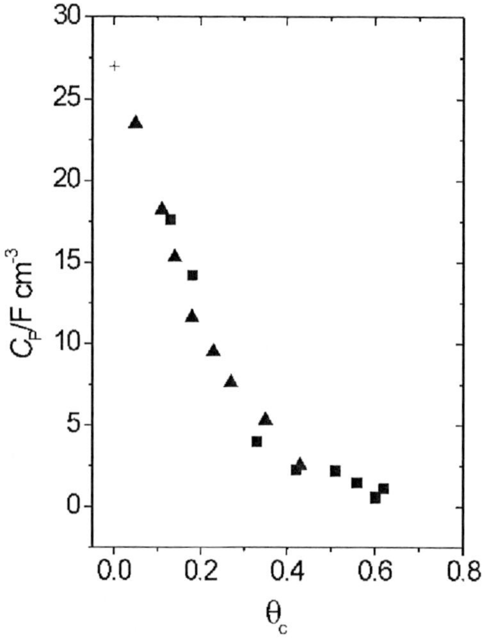

Figure 20. Redox capacitance (C_p) as a function of θ_c: Deactivated (■) and reactivated (▲) films. θ_c values for the 8 POAP films listed in Table 4. (+) Value corresponding to the freshly prepared POAP film. Electrolyte: 0.1 M $HClO_4$ + 0.4 M $NaClO_4$ + 2 x 10^{-3} M (HQ/Q) solution.

C_p *vs.* θ_c dependence is shown in Figure 20. As can be seen, starting from a C_p value of about 27 F cm^{-3}, for a freshly prepared film, a decrease of C_p with increasing θ_c is observed. The author of [53] indicates that there was no difference in the redox capacitance *vs.* θ_c dependence achieved in the supporting electrolyte from the -Z' *vs.* ω^{-1} slopes and in the presence of the redox active substrate from Eq. (5). Only small differences in the numerical values of C_p were found. The good agreement between the redox capacitance values attained under these different conditions was considered as an indication of the high fitting accuracy reached in [53] to obtain optimum parameter values (in this case C_p) for the treated system, by using Eq. (5).

As can be seen from Figure 21, $R_{m|f}$ exhibits an almost continuous increase from the beginning of the deactivation, that is, from θ_c = 0 to θ_c = 0.6. This increase was considered consistent with an increasing number of inactive sites at the metal|polymer interface, as the deactivation becomes more pronounced [50].

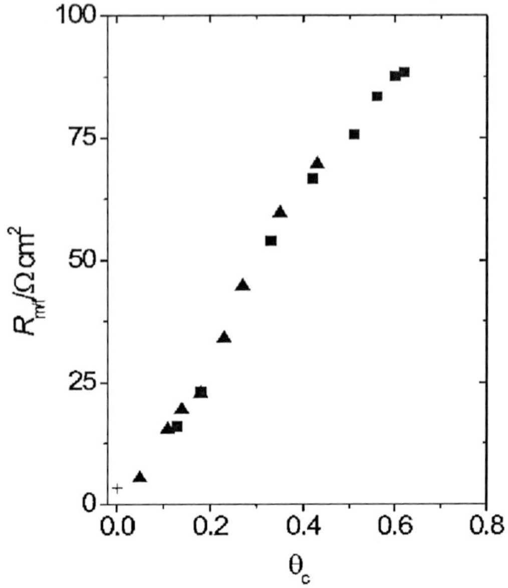

Figure 21. Metal-polymer interfacial electron-transfer resistance ($R_{m/f}$) as a function of θ_c. The same symbols and conditions as indicated in Figure 20.

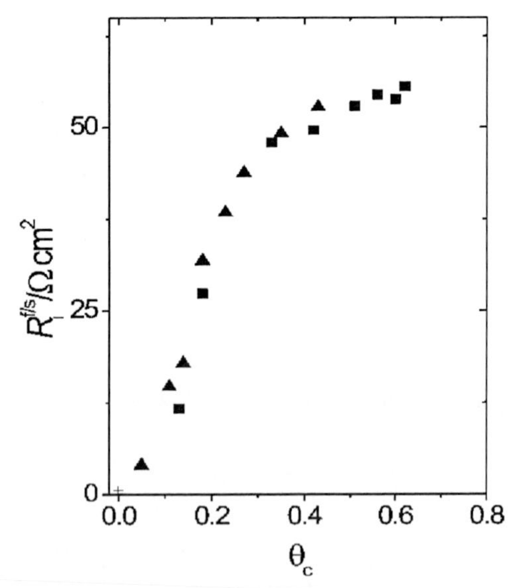

Figure 22. Polymer-solution interfacial ion-transfer resistance ($R_i^{f/s}$) as a function of θ_c. The same symbols and conditions as indicated in Figure 20.

$R_i^{f|s}$ as a function of θ_c is shown in Figure 22. This dependence exhibits a different feature as compared with $R_{m|f}$. That is, while a strong $R_i^{f|s}$ increase is observed within the range $0 < \theta_c < 0.35$, a less pronounced increase is obtained at values of $\theta_c > 0.4$.

As can be seen from Figure 23, within the whole range of θ_c values, the electron-transfer resistance $R_e^{f|s}$ at the polymer|solution interface shows a similar feature to that of $R_{m|f}$. However, $R_e^{f|s}$ results almost one order of magnitude lower than both $R_{m|f}$ and $R_i^{f|s}$. $R_e^{f|s}$ values were extracted from Eq. (7) using k_o as fitting parameter ($0.1 < k_o < 1$ cm s^{-1}). However, the author of [53] indicates that the accuracy of $R_e^{f|s}$ values is somewhat affected, especially at the beginning of the deactivation ($\theta_c < 0.2$) where the experimental situation $R_e^{f|s} << R_{m|f}$ and $R_i^{f|s}$ predominates. Then, $R_e^{f|s}$ values are more realistic at high degrees of deactivation ($\theta_c > 0.3$).

With regard to C_H values, Figure 24 shows that starting from a value of around 17 μF cm^{-2} for a freshly prepared film, C_H decreases in a nearly continuous way as the degree of deactivation increases, reaching a value of about 5 μF cm^{-2} for the maximum θ_c (~0.6) value. The C_H decrease, in the same way as the $R_{m|f}$ increase, was assigned to the creation of inactive gaps in the redox site configuration at the polymer|metal interface with deactivation.

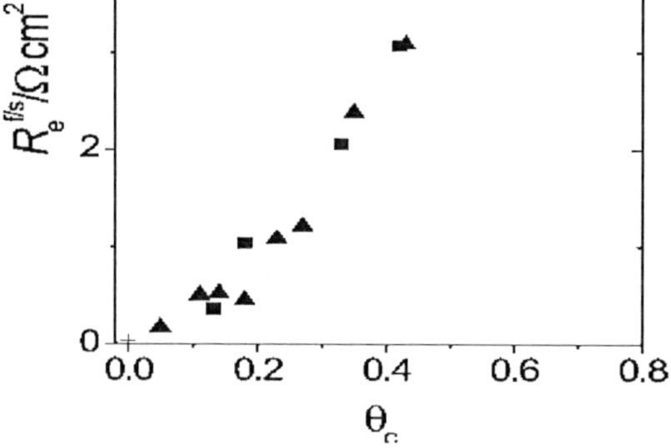

Figure 23. Interfacial electron-transfer resistance ($R_e^{f/s}$) as a function of θ_c. The same symbols and conditions as indicated in Figure 20.

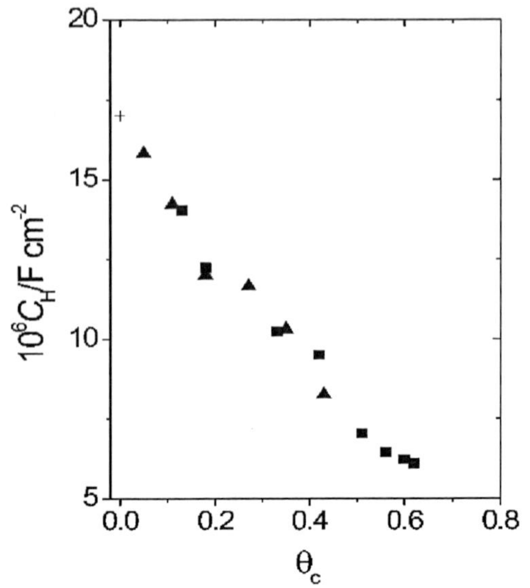

Figure 24. Interfacial capacitance (C_H) at the metal-polymer interface as a function of θ_c. The same symbols and conditions as indicated in Figure 20.

As can be seen from Figure 20 to Figure 24, when deactivated POAP films are reactivated by alkaline treatment, all dependences (C_p, $R_{m|f}$, $R_e^{f|s}$, $R_i^{f|s}$ and C_H *vs.* θ_c) shown in [53] do not exhibit hysteresis within the whole θ_c range. That is, all these dependences exhibit a univocal correspondence between deactivation and subsequent reactivation. However, this is not the case for the dependences that involve diffusion coefficients for electron (D_e) and ion (D_i) transport.

These dependences are shown in Figure 25 and Figure 26, respectively. As can be seen, even though both dependences seem to show a univocal correspondence between deactivation and subsequent reactivation for θ_c values lower than 0.4, a hysteresis phenomenon is observed after the reactivation of films with θ_c values higher than 0.5.

The univocal D_e *vs.* θ_c dependence within the deactivation range $0 < \theta_c < 0.4$ seems to be nearly linear with a slope $\Delta D_e/\Delta\theta_c = -1.6 \times 10^{-8}$ cm^2 s^{-1}. In this way, it was considered possible to recover the electron conductivity of a deactivated POAP film by reactivation within the θ_c value comprised between 0 and 0.4. The decrease of D_e with the increase of θ_c shown in Figure 25 was attributed to an increase of the hopping distance between remnant redox active sites after deactivation [52].

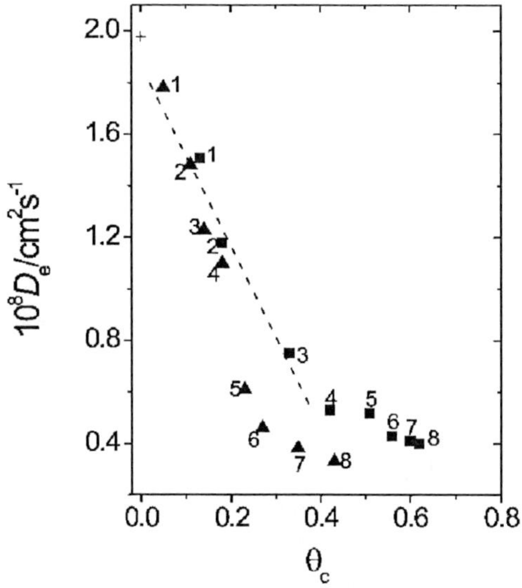

Figure 25. Electron diffusion coefficient (D_e) as a function of θ_c. Numbers 1 to 8 represent each one of the 8 films listed in Table 4. The same symbols and conditions as indicated in Figure 20.

The existence of an extended redox site configuration was assumed for a POAP film after deactivation, as compared with that corresponding to a freshly prepared one. In order to explain the effect of an extended redox sites configuration on the electron-transport process, the author of [53] invokes the expression of the electron-diffusion coefficient given in [52], $D_e = (a^2 k_o)$, where k_o is the intermolecular electron-transfer rate constant and a is the mean distance between two adjacent redox sites. The hopping rate, k_o, exhibits an exponential dependence on a, through the energy $-U(x+a)$ of a state with an electron in the position x along the current direction (see Eq. (23) in Ref. [52]). That is, the main effect of the hopping distance a on D_e proceeds from the exchange rate. In this sense, the effect of increasing the mutual separation a between adjacent redox centres is to decrease the exchange rate, k_o, which in turn should lead to a D_e decrease. The decrease of D_e with the increase of θ_c (Figure 25) seems to confirm results obtained by applying RDEV in [50], where the electron-transport rate represented by the electron current, I_e, also decreases as the degree of deactivation increases. However, despite the similarity between D_e *vs.* θ_c obtained in [53] and I_e vs. θ_c [50] dependences, it should be indicated that D_e values extracted from impedance measurements are almost two orders of magnitude higher than those obtained from RDEV.

However, it is well-known that the interpretation of diffusion coefficients at polymer films depends on the model employed to describe the charge propagation and sometimes on the technique used in its measurement. In this regard the rate of charge transport in electroactive polymers is usually characterized by an apparent diffusion coefficient, which can vary over several orders of magnitude (10^{-7}-10^{-14} $cm^2 s^{-1}$).

In terms of the electron-hopping model, the hysteresis phenomenon during the reactivation process was explained in terms of a lack of relaxation of the configuration of redox sites of a film, which was strongly deactivated ($\theta_c >$ 0.5). This lack of relaxation was attributed to the poor swelling effect at the reduced state of the polymer. In this sense, it was considered that at the reduced state of a strongly deactivated POAP film, the ability of the inner solution (incorporated into the polymer matrix) to reduce the electrostatic repulsion of the charged sites is limited. Under these conditions redox sites would adopt an extended configuration so as to minimize these coulombic repulsions, and then a more expanded redox site configuration could remain even after reactivation. With regard to ion transport, the D_i *vs.* θ_c dependence is shown in Figure 26.

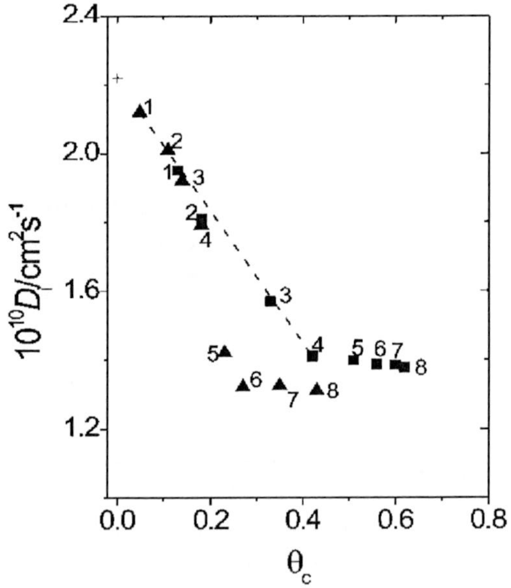

Figure 26. Ion diffusion coefficient (D_i) as a function of θ_c. Numbers 1 to 8 represent each one of the 8 films listed in Table 4. The same symbols and conditions as indicated in Figure 20.

This dependence exhibits the same characteristics as the D_e *vs.* θ_c dependence. That is, a univocal correspondence is observed between deactivation and reactivation, for θ_c values lower than 0.4 with a slope value of $\Delta D_i/\Delta\theta_c = - 2.0 \times 10^{-10}$ cm^2 s^{-1}. A change of slope is observed at about $\theta_c \sim 0.5$. A change of slope is also observed in the $R_i^{f|s}$ vs. θ_c dependence (Figure 22). The author of [53] associates both transport parameters $R_i^{f|s}$ and D_i with proton movements across the POAP|solution interface and inside the polymer film, respectively, rather than with anion transport.

To explain the observed ion diffusion coefficient change with the degree of deactivation, the existence of two forms (mobile and bound [31]) of hydrogen ions in the bulk film was proposed in [53]. In the regard, it was indicated that in polymers derived from aromatic amines, hydrogen ions could be constrained by nitrogen atoms of polymer chains and do not contribute to the electrical conductance of the film, and another part of such constrained groups is able to dissociate producing the mobile form of hydrogen, which provides the film conductance. That is, besides mobile protons, some traps for hydrogen ions within the bulk of the film may be present, which provides the binding of these protons with polymer film fragments. This proton conduction mechanism is similar to that proposed for ion transport in some solid-state materials [59], where ion movement occurs by a hopping process with the participation of two different types of sites. While some sites allow a fast ion diffusion process, other sites immobilize ions and do not allow ions to participate in the diffusion process. Thus, the pronounced initial increase of $R_i^{f|s}$ (and decrease of D_i) with deactivation within the range $0 < \theta_c < 0.4$ (Figure 22 and Figure 26, respectively) were attributed to the inhibition of nitrogen-containing groups that provide the binding of hydrogen ions and at the same time are able to dissociate and give the mobile form of hydrogen that contributes to the polymer conductivity. However, after more extensive deactivation ($\theta_c > 0.5$), it was assumed that inhibition of traps for hydrogen ions occurs and then, this only causes a small ion conductivity change (attenuated $R_i^{f|s}$ increase and D_i decrease, for $\theta_c > 0.5$). On the other hand, reactivation within the θ_c range comprised between 0 and 0.4 should allow recovering nitrogen-containing groups that provide the binding of hydrogen ions that contribute strongly to the film conductivity.

Then, diffusion coefficients of charge carriers into POAP films seem to be functions not only of the total amount of sites capable of participating in the charge-transport processes, that is, of the degree of deactivation, but also of the distribution of sites.

4. DEGRADED POAP FILM IN THE PRESENCE OF THE HQ/Q REDOX COUPLE

In practical uses of POAP, the applied potential plays a crucial role in the performance of the polymer. In this regard, it was demonstrated that POAP remains electroactive when the potential is cycled from -0.2 V to 0.5 V (*vs.* SCE). However, POAP undergoes irreversible deterioration when the positive potential limit is extended to values higher than a threshold value of 0.5 V (*vs.* SCE). This phenomenon was studied in [60] by employing CV, RDEV and EIS. CV was employed to prepare fresh POAP films and also to degrade them quantitatively. Then, the different behaviours of freshly prepared and degraded POAP films in contact with a solution containing the HQ/Q redox couple were compared by employing RDEV and EIS.

The resulting experimental impedance diagrams were also interpreted on the basis of the same homogeneous impedance model described in [15]. The different dependences of the charge-transfer and charge-transport parameters on the degree of degradation of the polymer ($\theta_{Red,Max}$) are shown in [60]. The different features of some of these dependences (for instance, bulk electronic and ionic transport on the degree of deactivation) were explained in terms of the different mechanisms of electron and ion transport in polymeric materials. The charge-transfer resistance at the metal|polymer interface, $R_{m|f}$, exhibits a strong increase from the beginning of the degradation.

At the same $\theta_{Red,Max}$ values, the ion-transfer resistance $R_i^{f|s}$ at the polymer|solution interface is almost one order of magnitude lower than $R_{m|f}$. Also, a pronounced increase in the resistance related to the electron interfacial exchange $R_e^{f|s}$ at the polymer|solution interface was observed from the beginning of the polymer degradation. However, $R_e^{f|s}$ values are nearly one order of magnitude lower than $R_{m|f}$ values. Then, concerning the polymer|electroactive solution interface, a slow ion-transfer process was observed for POAP as compared with the electron-transfer process at the metal|polymer interface. Electron (D_e) and ion (D_i) diffusion coefficient changes with $\theta_{Red,Max}$ were also extracted from the impedance model described in [15].

An interesting comparison was established in [60] between the transversal charge-transfer resistance at the metal|polymer interface $R_{m|f}$ (obtained by impedance spectroscopy) and the lateral resistance $\Delta R/R$ along the electrode (obtained by surface resistance measurements [61]), as a function of $\theta_{Red,Max}$.

CONCLUSION

Poly(*o*-aminophenol) has received growing attention due to its potential application in microelectronic devices, analytical sensors, and electrochromic displays. However, from the practical viewpoint it is necessary that the conducting properties of the films remain substantially unchanged, before and after the polymer has been used. Results described in [50,53,60] demonstrate that when poly(*o*-aminophenol) film electrodes are subjected to rough conditions (excessive positive potentials and contact with inhibitors of redox centres) as in some of their practical applications, the films can be deactivated or even degraded. While for low degrees of deactivation (lower than 0.4) it is possible to recover the conductivity of a deactivated POAP film by treatment with an alkaline solution (reactivation), the degradation caused by excessive positive potential limits leads to an irreversible deterioration of the polymer film. These deactivation and degradation phenomena were demonstrated by studying the electrochemistry of poly(*o*-aminophenol) films in the presence of the hydroquinone/*p*-benzoquinone redox couple. The neutral hydroquinone/*p*-benzoquinone redox couple exhibits some advantages over other charged redox couples employed to study conduction processes at polymer films. Redox couples containing charged species can form ion pairing with the polymer matrix, which restricts their diffusion rates during permeation processes. Diffusion rate values for neutral hydroquinone and *p*-benzoquinone species across poly(*o*-aminophenol) films were determined at different film thicknesses and external solution compositions. Even thought our poly(*o*-aminophenol) films were synthesized and acidic media, the study of the diffusion processes of electroactive species across the films carried out in our laboratory could be interesting from a comparative viewpoint. In this regard, poly(*o*-aminophenol) synthesized at *pH* values over 3 has usefully been employed in the development of different types of sensors, mainly biosensors based on immobilized enzymes. In these biosensors the poly(*o*-aminophenol) film acts as an effective barrier to protect the sensor from the interference of electroactive species. In this regard, the low permselectivity of the polymer was fund to be suitable to reduce the effect of interferents, such as ascorbic acid, uric acid and acetaminophen, in different biosensors. The redox mediation process at the poly(*o*-aminophenol)|hydroquinone/*p*-benzoquinone interface allowed demonstrating that proton and electron diffusion inside the film and rates of interfacial charge-transfer processes are strongly reduced during deactivation and degradation of the films. It should be expected that these deterioration processes drastically reduce the efficiency of the material in its practical applications. This review work can help gain further insights into the charge-propagation process at poly(*o*-aminophenol), in such a way that the restrictive stability conditions of the polymer in its practical applications can be exploited more successfully.

ACKNOWLEDGMENTS

The authors gratefully acknowledge the Consejo Nacional de Investigaciones Científicas y Técnicas (CONICET) and also the Facultad de Ciencias Exactas, National University of La Plata (UNLP).

REFERENCES

[1] Linford, R.G. (Ed.), *Electrochemical Science and Technology of Polymers,* 1987, Elsevier, England, vol. 1.
[2] Linford, R.G. (Ed.), *Electrochemical Science and Technology of Polymers,* 1990, Elsevier, England, vol. 2.
[3] Bruce, P.J. *Solid State Electrochemistry*, 1995, Cambridge, University Press, Cambridge.
[4] Lyons, M.E.G. (Ed.), *Electroactive Polymer Electrochemistry*, 1994, Plenum Press, New York, Part 1.
[5] Lyons, M.E.G. (Ed.), *Electroactive Polymer Electrochemistry*, 1996, Plenum Press, New York, Part 2.
[6] Evans, G.P. *Electrochemical Science Engineering*, 1990, Eds: Gerischer, H.; Tobias,C.W. vol. 1, VCH Press, Wenheim, p 1.
[7] Murray, R.W., *Electroanalytical Chemistry*, 1984, Ed.: Bard, A.J., Marcel Dekker, New York, Vol. 13, pp 191-197.
[8] Skotheim, T.A. (Ed.), *Handbook of Conducting Polymers*, 1966, Marcel Dekker, New York.
[9] Kuzmany, H.; Mehring, M.; Roth, S. (Eds), *Electronic Properties of Conjugated Polymers,* 1997, Springer-Verlag, Berlin.
[10] Inzelt, G.; Pineri, M.; Schultze, J.W.; Vorotyntsev, M.A. *Electrochimica Acta.* 2000, 45, 2403-2421.
[11] Malev, V.V.; Kondratiev, V.V. *Russian Chemical Reviews*, 2006, 75, 147-160.

[12] Anderson, J.L.; Bowden, E.F.; Pickup, P.G. *Analytical Chemistry*, 1996, 68, 379R-444R.
[13] Ybarra, G.; Moina, C.; Florit, M.I., Posadas, D. Electroanaytical Chemistry, 2007, 609, 129-139.
[14] Ybarra, G.; Moina, C.; Florit, M.I., Posadas, D., *Electrochimical Acta,* 2008, 53, 3955-3959.
[15] Vorotyntsev, M.A.; Deslouis, C.; Musiani, M.M.; Tribollet, B.; Aoki, K. *Electrochimica Acta*, 1999, 44, 2105-2115.
[16] Vorotyntsev, M.A. *Electrochimica Acta*, 2002, 47, 2071-2079.
[17] Barbero, C.; Silber, J.J.; Sereno, L. *J. Electroanal. Chem.* 1989, 263, 333-352.
[18] Barbero, C.; Silber, J.J.; Sereno, L. *J. Electroanal. Chem.* 1990, 291, 81-101.
[19] Ohsaka, T.; Kunimura, S.; Oyama, N. *Electrochim. Acta*, 1988, 33, 639-645.
[20] Guenbourg, A.; Kacemi, A.; Benbachir, A.; Aries, L. *Prog. Org. Coat.*, 2000, 38, 121-127.
[21] Kunimura, S.; Ohsaka, T.; Oyama, N. *Macromolecules,* 1988, 21, 894-900.
[22] Yang, Y.; Lin, Z. *Synth. Met.*, 1996, 78, 111-115.
[23] Barbero, C.; Tucceri, R.I.; Posadas, D.; Silber, J.J.; Sereno, L. *Electrochim. Acta*, 1995, 40, 1037-1040.
[24] Tucceri, R.I.; Barbero, C.; Silber, J.J.; Sereno, L.; Posadas, D. *Electrochim. Acta*, 1997, 42, 919-927.
[25] Komura, T.; Ito, Y.; Yamaguti, T.; Takahasi, K. *Electrochim. Acta,* 1998, 43, 723-731.
[26] Ortega, J.M. *Thin Solid Films*, 2000, 371, 28-35.
[27] Tucceri, R.I. *J. Electroanal. Chem.*, 2001, 505, 72-84.
[28] Tucceri, R.I. *J. Electroanal. Chem.*, 2003, 543, 61-71.
[29] Salavagione, H.J.; Arias, J.; Garcés, P.; Morallón, E.; Barbero, C.; Vázquez, J.L. *J. Electroanal Chem.*, 2004, 565, 375-383.
[30] Salavagione, H.J.; Arias-Padilla, J.; Pérez, J.M.; Vázquez, J.L.; Morallón, E.; Miras,M.C.; Barbero, C. *J. Electroanal Chem.*, 2005, 576, 139-145.
[31] Levin, O.; Kondratiev, V.; Malev, V. *Electrochim. Acta,* 2005, 50, 1573-1585.
[32] Pan, D.; Chen, J.; Nie, L.; Tao, W.; Yao, S. *Electrochim. Acta*, 2004, 49, 795-801.

[33] Pan, D.; Chen, J.; Yao,S.; Tao, W.; Nie, L. *Analytical Sciences*, 2005, 21, 367.

[34] Chen, X.; Chen, J.; Deng, C.; Xiao, C.: Yang, Y.; Nie, Z.; Yao, S. *Talanta,* 2008, 76, 763-767.

[35] Li, J.; Lin, X. *Biosensors and Bioelectronics*, 2007, 22, 2898-2905.

[36] Miland, E.; Miranda Ordieres, A.J.; Tuñón Blanco, P.; Smyth, M.R.; Fágáin, C.O. *Talanta,* 1996, 43, 785.

[37] Lobo Castañón, J.; Miranda Ordieres, A.J.; Tuñón Blanco, P. *Anal. Chim. Acta,* 1997, 346, 165-174.

[38] Lobo, M.J.; Miranda, A.J.; López-Fonseca, J.M.; Tuñón, P. *Analytica Chimica Acta,* 1996, 325, 33-42.

[39] Valdés García, M.A.; Tuñón Blanco, P.; Ivaska, A. *Electrochim. Acta,* 1998, 43, 3533-3539.

[40] Zhang, Z.; Liu, H.; Deng, J. *J. Anal. Chem.,* 1996, 68, 1632-1638.

[41] Golabi, M.S.; Nozad, A. *Electroanalysis,* 2003, 15, 278-286.

[42] Zhang, A.Q.; Cui, C.Q.; Lee, J.Y. *J. Electroanal. Chem*., 1996, 413, 143-151.

[43] Yano, J.; Kawakami, H.; Yamasaki, S.; Kanno, Y. *Electrochemical Society.,* 2001, 148, E61-E65.

[44] Miras, M.C.; Badano, A.; Bruno, M.M.; Barbero, C. Portugaliae *Electrochimica Acta,* 2003, 21, 235-243.

[45] Bonfranceschi, A.; Pérez Córdoba, A.; Keunchkarian, S.; Zapata, S.; R. Tucceri, R. *Electroanalytical Chemistry*, 1999, 477, 1-13.

[46] Ikeda, T.; Schmehl, R.; Denisevich, P.; Willman, K.; Murray, R.W. *J. Am. Chem. Soc.,* 1982, 104, 2683.

[47] Soto, J.E.; Kim, Y.G.; Soriaga, M.P. *Electrochem. Commun.* 1999, 1, 1135.

[48] Rodríguez Nieto, F.J.; Posadas, D.; Tucceri, R. *Electroanalalytical Chemistry,* 1997, 434, 83-91.

[49] Deslouis, C.; Tribollet, B. *Advances in Electrochemical Science and Engineering*, 1992, Eds.: Gerischer, H.; Tobias, C., VCH Publishers, Vol. 2, New York, USA.

[50] Tucceri, R.I. *Electroanalytical Chemistry*, 2009, 633, 198-206.

[51] Rodríguez Nieto, F.J.; Tucceri, R. *Electroanalytical Chemistry*, 1996, 416, 1-24.

[52] Chidey, Ch.E.D.; Murray, R.W. *J. Phys. Chem*., 1986, 90, 1479-1484.

[53] Tucceri, R.I. *Electroanalytical Chemistry*, 2011, 659, 83-91.

[54] Musiani, M.M. *Electrochimimica Acta*, 1990, 35, 1665-1670.

[55] Agarwal, P.; Orazem, M.E.; García-Rubio, L.H. *Electrochemical Society,* 1992, 139, 1917-1927.
[56] Agarwal, P.; Crisalle, O.D.; Orazem, M.E.; García-Rubio, L.H. *Electrochemical Society*, 1995, 142, 4149-4158.
[57] Agarwal, P.; Orazem, M.E.; García-Rubio, L.H. *Electrochemical Society*, 1995, 142, 4159-4168.
[58] Orazem, M.E., *Electroanalytical Chemistry*, 2004, 572, 317-327.
[59] Bisquert, J. *Electrochimica Acta*, 2002, 47, 2435-2449.
[60] Tucceri, R. *Journal of New Materials for Electrochemical Systems,* 2005, 8, 305-315.
[61] Tucceri, R. *Surface Science Reports*, 2004, 56, 85-157.

In: Hydroquinone
Eds: F. Gokden and A. Lazzarotto

ISBN: 978-1-62100-258-1

Chapter 4

DETECTION OF THIOLS USING HYDROQUINONE ON GOLD SURFACE

Luisa Rojas de Astudillo,[1] Rosa Brito Gómez[2] and Rolando J. Tremont[3*]

[1]Departamento de Química, Universidad de Oriente, Avenida Universidad, Cumaná-Sucre, Venezuela
[2]Department of Natural Science, Interamerican University of PR, Metropolitan Campus, Puerto Rico
[3]Department of Chemistry, University of Puerto Rico at Humacao, CUH Station, Humacao, Puerto Rico

ABSTRACT

Benzene, declared as an environmental contaminant, has commonly been used as an industrial solvent, a starting material for the synthesis of other chemicals, and as an antiknock agent in gasoline, and is a significant component of cigarette smoke. While not fully understood, the toxicity of benzene is linked to the production of reactive metabolites formed during cytochrome P450 2E1. Benzene is metabolized by cytochrome P450 CYP2E1 to benzene oxide, which spontaneously forms phenol. Phenol is converted to hydroquinone by P4502E1. Hydroquinone and related metabolites are converted in the bone marrow, by

[*] Department of Chemistry, University of Puerto Rico at Humacao, CUH Station, 100 Road 908, Humacao, P. R. 00791-4300. E-mail: rolando.tremont@upr.edu.

myeloperoxidase, to benzoquinones, which are potent hematotoxic and genotoxic compounds that undergo further metabolic activation, by one-electron reducing enzymes, to generate semiquinones and reactive oxygen species (ROS). As an excess of ROS is produced, oxidative stress can occur and lead to altered cellular signaling. The production of ROS is tightly regulated and the cell possesses detoxifying mechanisms such as catalase, glutathione and superoxide dismutase to maintain appropriate cellular ROS levels. 1,4 benzoquinone (BQ) can alter signaling pathways and affect chromosomal integrity. Numerous attempts have been made to understand cytotoxic mechanisms influenced by the chemical structure of BQ generate DNA breaks and apoptosis. It is also a strong topoisomerase II poison. Its toxicity is mainly due to oxidative stress and/or forming Michael-type adducts with proteins and glutathione (GSH). Owing to the important physiological roles of quinones, such as hydroquinone, 1,4 benzoquinone, and reduced thiols such as GSH, it is necessary to develop sensitive and selective methods for the measurement of those substances to study cellular mechanisms of chemical-induced toxicity. For this purpose, the benzoquinone (BQ) is used in the electrochemical determination of thiol molecules. The Addition of sulfur nucleophiles to quinones has been studied and Characterized as 1, 4-reductive addition of the Michael type. Quinones, containing a polarized double bond, are considered Michael acceptors, where the Michael donors are thiol nucleophiles. The final product of this reaction is hydroquinone with a sulfur atom in the ortho position. If a metal surface is modified with a dithiol (to form a thiolate with the metal surface, leaving a free thiol,-HS) molecule, the reaction with BQ will be easy to produce hydroquinone (H2Q) moiety. The latter product can be used to determine electrochemically thiol-containing compounds in its structure.

Keywords: 2-propene-1-thiol, 1,3-propanedithiol, 1,4-benzoquinone, hydroquinone, XPS

INTRODUCTION

Thiols play an important role in metabolic processes of all living cells and, take part in a wide variety of intracellular oxidation–reduction reactions [1,2]. These compounds are commonly known as mercaptans, a class of organic compounds containing sulphydryl groups (–SH) attached to a carbon atom [3]. Among mercaptans, glutathione (GSH), cysteine (Cys), homocysteine (Hcys), cysteinylglycine and γ-glutamylcysteine are the most important endogenous aminothiols in human biofluids [4]. The presence of N-acetylcysteine and thioglycolic acid in urine has been confirmed and these compounds are assumed to be endogenous constituents of human urine [5, 6].

GSH, discovered in 1888, is a ubiquitous thiol-containing tripeptide, and the major component of cellular antioxidant defenses [7, 8], protects cells from oxidative stress and nitrosative damage [9]. As a key modulator of cell functions has important roles in redox regulation of proteins and maintaining redox homeostasis that is critical for proper function of cellular processes, including apoptosis [10]. Its levels in human tissues normally range from 0.1 to 10 mM with high concentrations in liver (up to 10 mM) and in the spleen, kidney, lens, erythrocytes and leukocytes [11]. The GSH system, present in both cytoplasm and mitochondria, provides most of the H_2O_2 removing capacity in majority of cells [12].

Under oxidative conditions GSH is oxidized to its disulfide (GSSG) and exported from cells when its concentration increases. The decrease in GSH and the concomitant decrease in the GSH/GSSG ratio have been suggested to be important contributing factors to some human diseases such as lung inflammation, amyotrophiclateralsclerosis, chronic renal failure, malignant disorders, diabetes, Parkinson's disease, Alzheimer's disease, catar act formation as well as to the ageing process [13,14,15].

In children with severe malnutrition, it has consistently been shown that the concentration of GSH in plasma and whole blood is lower than normal in those with edema but not in those without edema [16,17]. These results point out that a lower concentration of GSH is indicative of impaired antioxidant capacity [7].

Compared with GSH, Cys is a very abundant extracellular aminothiol [7] that is found in large amounts (0.5–10 mM) inside the cell [18]. This amino acid participates in a number of biochemical processes, and is involved in several important cellular functions, including protein synthesis, detoxification and metabolism [19]. It is crucial to determine the concentrations of Cys in plasma and urine, because elevated level of Cys in plasma is considered as a vascular disease risk factor [20-22] and is associated with neurotoxicity [23]. On the contrary, deficiency of Cys may result in some serious diseases such as hematopoiesis decrease, leucocyte loss, and psoriasis [24]. Moreover, altered level of Cys has been implicated in hyperhomocysteinemia [25], which has been linked to the increased risks of Alzheimer's disease [26], neural tube defect [27], and osteoporosis [28]. Hence, the detection of Cys continues to be of interest [29].

During the biosynthesis of the essential amino acid methionine to cysteine, homocysteine (Hcy) is formed as an intermediate, which is at the intersection of two main metabolic pathways: remethylation and trans-sulfuration, regulated by methylenetetrahydrofolate reductase and cystathionine beta-synthase [30]. Several studies have revealed that elevated levels of total homocysteine in plasma in fasting is associated with an increased risk for atherosclerotic and tromboembolic vascular disease [31,32]. One of the major mechanisms of Hcy toxicity seemed to be auto-oxidation of Hcy, which reduces disulfide to free thiol, followed by the metalindependent oxidation of the free thiol to generate oxidants such as superoxide and hydrogen peroxide [33].

Other thiols commonly used as drugs in the treatment of many diseases are Cysteamine, N-acetylcysteine, captopril, mesna, 2- mercaptopropionylglycine, D-penicillamine, thyreostats and thiopurines. However, severe adverse reactions to oral thiol-drugs use have been described in subjects in which abrupt incremental dosing of the drugs was started [4].

The information mentioned above suggests that there is a need for the development of simple yet selective protocols for thiol determination to implement preventative actions before the manifestation of physical symptoms. The classical method currently used to quantify thiols is the Ellman's test, which is a spectrophotometric method based on an exchange reaction between the thiol and the disulfide DTNB (5,50-dithiobis-(2-nitrobenzoic acid)). The sulfide produced shows a characteristic yellow color. However, this test and more generally spectrophotometric methods require a pre-treatment when thiols have to be assayed in coloured samples and their results may be influenced by variable levels of specific enzyme activities such as glutathione S-transferase or g-glutamyltransferase [34].

Other intensive studies have been evaluated to the detection of biological thiols, more particularly glutathione. Among them, several quinone systems have been investigated as potential indicators for reduced thiol species [35,36]. The addition of thiols to quinone moieties represents the basis of numerous spectroscopic [37, 38] and pre-column chromatographic derivatization protocols [39,36].

The analytical community has focused its attention on the redox properties of the aminothiols Hcy, Cys, and GSH [35,40,41]. Electrochemical techniques hold much potential for the in vitro analysis of these aminothiols because such methods offer the advantage of minimal sample pretreatment, rapid analysis time, and simple experimental approach. Numerous electroanalytical strategies have been employed to aid the quantification of thiols and various merits and limitations of the approaches have been reviewed [42, 40, 34].

Benzoquinone (BQ) is also used in the electrochemical determination of thiol molecules because of the interaction of the hydroquinone/benzoquinone redox couple with thiols, thus benzoquinone behaves as an electron mediator for thiols detection in the electroanalytical approach [35, 43]. The addition of sulfur nucleophiles to quinones has been studied and characterized as 1, 4-reductive addition of the Michael type (Eq. 1). Quinones, containing a polarized double bond, are considered Michael acceptors, where the Michael donors are thiol nucleophiles. The final product of this reaction is hydroquinone with a sulfur atom in the ortho position. If a metal surface is modified with a dithiol (to form a thiolate with the metal surface, leaving a free thiol,-HS) molecule, the reaction with BQ will be easy to monitor through the production of hydroquinone (H_2Q) moiety [44, 45].

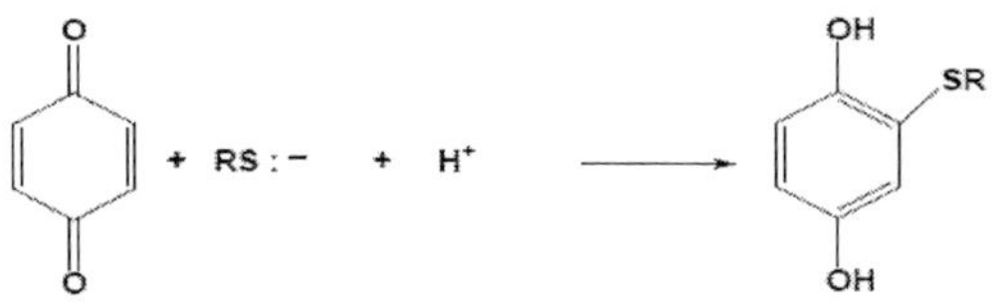

(Eq. 1)

The product orientation in the addition of sulfur nucleophiles to quinones is determined by the substituent and the occurrence of the new sulfur substituent controls in large extent the subsequent chemistry.

The reduction potential of quinoid compounds is, among the intrinsical properties of quinones, the most important factor affecting the rate of electron transfer between donor and acceptor. Moreover, the reduction potential represents the simplest control over free radical reactivity [46].

Because quinones are one- and two-electron redox species, following reduction potentials at pH 7 can be considered: the one-electron reduction potential [$E(BQ/BQ\bullet^-)$], the two-electron reduction potential [$E(BQ/BQ^{2-})$], and the reduction potential of the intermediate step [$E(BQ\bullet^-/BQ^{2-})$]. The latter, $E(BQ\bullet^-/BQ^{2-})$, is not easily measured in protic media but it can be calculated according to the formula [47]:

$$E(BQ\bullet^-/BQ^{2-}) = 2\,E(BQ/BQ^{2-}) - E(BQ/BQ\bullet^-)$$

The reduction potential of quinones is influenced by the substituents, dissociation constants, H^+ concentration, and solvent effects [48, 49]. The redox potential of quinones is dependent on the H^+ concentration of the medium and the dissociation constants of the species. At pH 7, semiquinones, having pK values between 4 and 5, will be in their anionic form and participate readily in electron transfer reactions; the so-called stability of hydroquinones, however, will be partly due to the fact that they are protonated at this pH, hence slowing down electron-transfer processes [50].

The latter product can be used to determine electrochemically thiol-containing compounds in its structure [51, 52]. Moreover, the gold electrodes have demonstrated to provide a suitable support for the thiol immobilization, the chemical modification of surface can be carried out with a very good reproducibility, suitable for the sensor development [53].

The electrochemical kinetics of the BQ/H_2Q redox couple in aqueous solutions has been the subject of numerous investigations [54-61]. In the electrochemical determination of thiols, by their ability to operate as biomarkers, BQ plays an important role in these studies. White et al. [62] studied different quinoid intermediates and their subsequent reaction with sulphydryl thiols (RSH), using glassy carbon as the working electrode. Others researcher has studied potentiometric response of the reaction of BQ and thiols [63-67]. Self-assembled monolayers (SAMs) are ordered molecular assemblies formed by the adsorption of active surfactants on a solid surface. This is a technique that provides an elegant route to the preparation of well-defined organic assemblies on solid surfaces [68]. The preparation of high-ordered monolayer by molecular self-assembly has been employed extensively as a surface derivatization procedure. This method has a wide use in the preparation of modified surfaces [68-97]. The advantages of SAMs include their ease of preparation, stability, and the possibility of introducing different chemical functionalities. The incorporation of the appropriate chemical functionality with some molecular level control into the highly ordered monolayer allows the preparation of surfaces with tailor-made properties [98]. The formation of high order monolayers is often done by the spontaneous adsorption of n-alkanethiols or their derivatives to gold surfaces [99-105].

In our laboratory was studied the modification of the gold surface with 1-propanethiol (1-PT), 2-propene-1-thiol (2-P-1-T) and 1,3-propanedithiol (1,3-PDT), leaving the thiol (-SH) group on the surface (see Figure 1 and scheme 1). After, it is putted to react the 1,4-benzoquinone (BQ) with the thiol (-SH) from surface, by Michael addition reaction, forming the hydroquinone (H_2Q) moiety. Here is presented results of the electrochemical behavior, using cyclic voltammetry, and a study of surface analysis, using X-ray photoelectron spectroscopy (XPS). For this chapter is of great interest to contribute with new modifications of gold surfaces and their possible applications in the field of biomarkers.

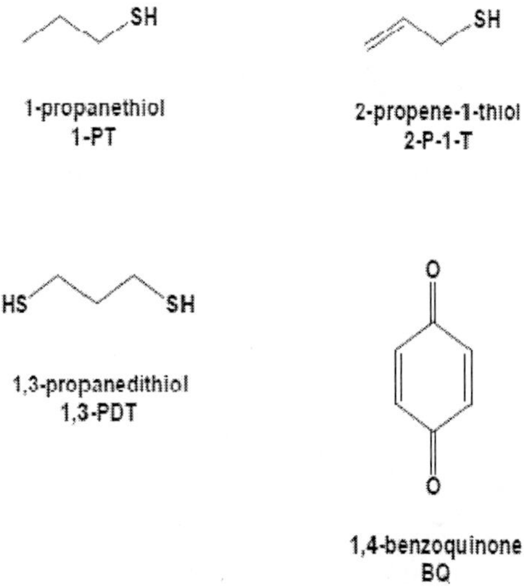

Figure 1. Structures of compounds used to modify gold surface.

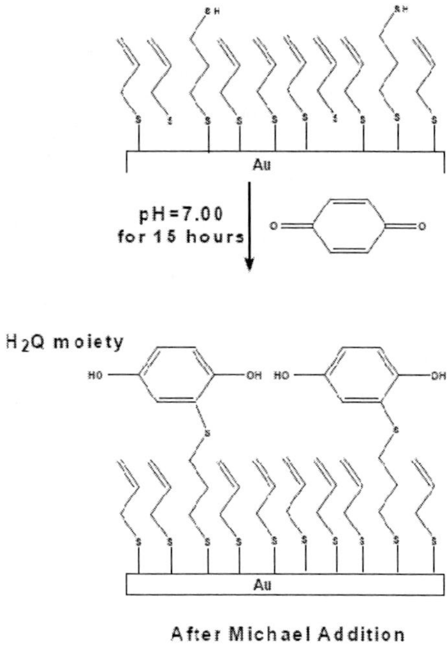

Scheme 1.

Electrochemical and XPS Studys of Gold Surface Modified with BQ

Pre-Treatment and Substrates Modification for Electrochemical Study

The electrochemical cell employed was a conventional three-electrode cell using platinum gauze as a counter electrode. All potentials are reported with respect to Ag/AgCl reference electrode. All substrates used were new and before their modification they were submitted to a cleaning treatment. Electrode cleanness was verified by cycling the potential in a 1.0 M H_2SO_4 solution between hydrogen and oxygen regions until characteristic cyclic voltammogram (CV) were observed. Was verified its cleanness by cyclic voltammetry before being chemically modified. Figure 2 shows a cyclic voltammogram of a clean gold electrode of 1.6 mm of diameter in 1.0 M H_2SO_4, at 100 mV/s scan rate. The current-potential behavior shown in this figure is characteristic of a clean Au surface in a clean test solution [106-109].

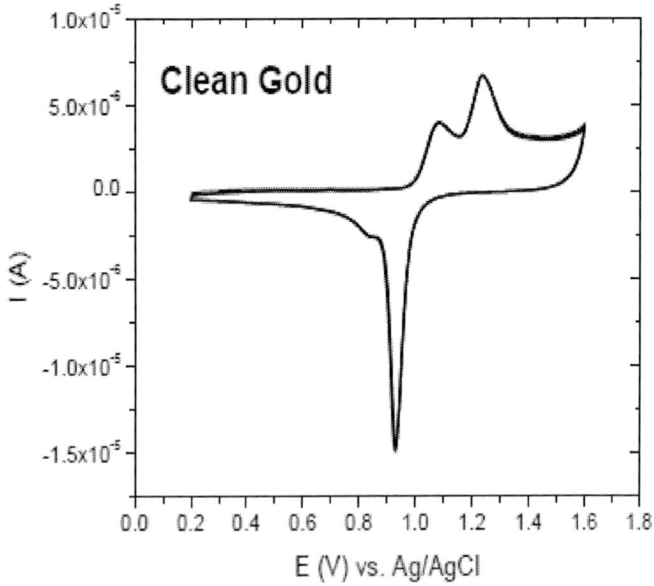

Figure 2. Cyclic voltammogram of clean gold surface in 1.0 M H_2SO_4 solution. Scan rate 100 mV s^{-1}. The solution was degassed with N_2 flow for least 30 min. Measurements were performed at 21 $\pm$ 1 °C.

Self-assembly of 1-PT (or 2-P-1-T) and 1, 3-PDT was done on Au surfaces using commercial gold disk electrodes of 1.6 mm of diameter to carry out the electrochemical measurements.

The previously cleaned substrates were submerged in a 3.0 mM 1-PT (or 2-P-1-T) ethanol solution for 30 min. afterward, the substrates were washed with ethanol, dried under nitrogen gas flow and submerged in a 3.0 mM 1, 3-PDT ethanol solution for 5 h.

The modified electrodes were immersed in a 2.01 mM 1, 4-benzoquinone (BQ) solution, in Britton-Robinson buffer (acetic, boric and phosphoric acids, each at a concentration of 0.04 M and adjusted to pH 7 through the addition of sodium hydroxide) [110], for 15 h. In this last solution, the BQ compound react through a nucleophilic 1,4-Michael addition to form the corresponding reduced thiol conjugate on gold surface (shown in scheme 1), forming the hydroquinone (H_2Q) moiety.

Finally, the electrochemical study of this latter system formed on the gold surface, is done in Britton-Robinson buffer solution without the presence of BQ in the solution.

Results of Cyclic Voltammetry Studies

The electrochemical characterization of the modified gold surfaces can reveal the presence of a full coverage of the electrode. Figure 3 present a cyclic voltammogram (CV) in 2.5 mM $K_3Fe(CN)_6$ in 0.1 M KCl solution: (a) for clean gold, (b) gold surface modified with 3.0 mM 2-P-1-T for 30 min and (c) gold surface modified with 3.0 mM 2-P-1-T for 30 min and 3.0 mM 1,3-PDT for 5 h (c), in alcoholic solution. Is observed that increasing the amount of organic molecules on the gold surface the reduction and oxidation current of ferricyanide species become smaller, indicating a blockade of the surface.

Using small molecules containing a thiol (-SH) group facilitates the transfer of electrons from the metal surface to the active group that is on this surface. Furthermore, if these molecules are unsaturated with one double bond (C=C), promotes electron transfer, which was demonstrated in this study.

Is placed the 1-PT (or 2-P-1-T) molecule on the gold surface to create pinholes between these molecules and then arranged inside these pinholes the 1,3-PDT molecule, leaving a thiol group (-SH) on the surface, which reacts with the BQ to form hydroquinone moiety, staying anchored to the surface (see scheme 1). This system would be used to recognize molecules containing a thiol (-SH) group in its structure, with an activity electrochemistry.

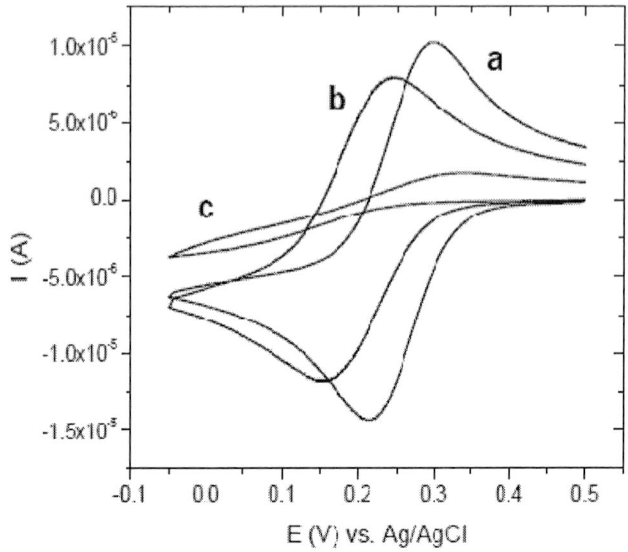

Figure 3a. Cyclic voltammograms of 2.5 mM $K_3Fe(CN)_6$ in 0.1 M KCl solution: (a) for clean gold, (b) gold surface modified with 3.0 mM 2-P-1-T for 30 min and (c) gold surface modified with 3.0 mM 2-P-1-T for 30 min and 3.0 mM 1,3-PDT for 5 h (d), in alcoholic solution. Scan rate 100 mV s^{-1}. Measurements were performed at 21 $\pm$ 1 °C.

Figure 3 shown voltammograms for a 2.01 mM 1,4-benzoquinone (BQ) solution, in Britton-Robinson buffer, at pH 7, on clean gold surface. As is obvious, in this potential window, in figure 3(b), a peak of reduction (E_{pc}) of +0.008 V and peak of oxidation (E_{pa}) of +0.121 V, with a peak-to-peak separation (ΔE_p) of 0.112 V, were observed. This behavior may be related to the one-electron redox potential [E(BQ/BQ•⁻)], with kinetic limitation. An anodic peak current (i_{pa}) of 16.4 μA and a cathodic peak current (i_{pc}) of -21.4 μA, for a $|i_{pa} / i_{pc}|$ of 0.76, describing this system as quasi-reversible electrochemically, also were observed. The BQ is reduced on clean gold surface, but in their oxidation chemical reactions coupled to the electrode process perturb the electron transfer. Gold surfaces were modified with 1-PT, 2-P-1-T, and 1,3-PDT. The gold modified surfaces were immersed in 2.01 mM 1,4-benzoquinone (BQ) solution, in Britton-Robinson buffer, at pH 7, for 15 h, without applying any potential. Afterward, the gold modified surfaces dropped out of nucleophile solution, washed with water, and immersed in Britton-Robinson buffer without BQ molecules. In this process, the BQ react with the thiol (-SH) group of the 1,3-PDT molecules, to form hydroquinone (H_2Q) moiety (see scheme 1), after Michael addition.

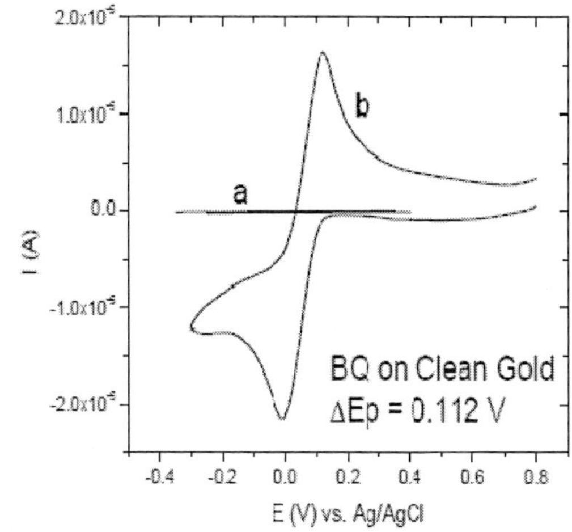

Figure 3b. Cyclic voltammograms for a 2.01 mM 1, 4-benzoquinone (BQ) solution, in Britton-Robinson buffer, at pH 7, on clean gold surface. The solution was degassed with N_2 flow for least 30 min. Scan rate 100 mV s^{-1}. Measurements were performed at 21 ± 1 °C.

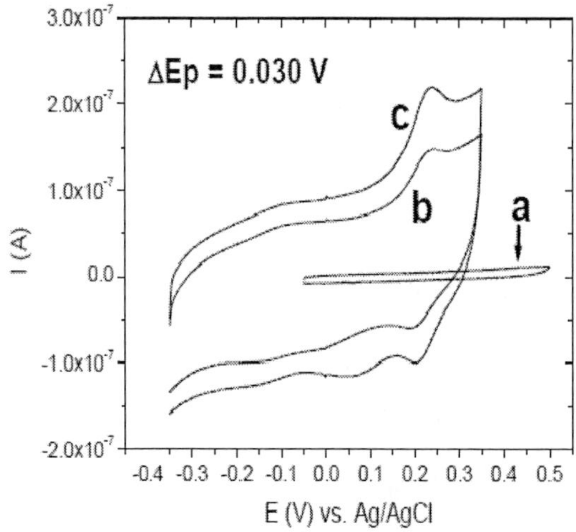

Figure 4. Cyclic voltammograms for the gold modified with: (a) 3.0 mM 2-P-1-T for 30 min and 3.0 mM 1,3-PDT for 5 h; (b) 3.0 mM 1-PT for 30 min, 3.0 mM 1,3-PDT for 5 h, and hydroquinone (H_2Q) moiety; (c) 2-P-1-T for 30 min, 3.0 mM 1,3-PDT for 5 h and hydroquinone (H_2Q) moiety. The solution was degassed with N_2 flow for least 30 min. Scan rate 100 mV s^{-1}. Measurements were performed at 21 ± 1 °C.

Figure 4 shown the CVs for the gold modified with (a) 3.0 mM 2-P-1-T for 30 min and 3.0 mM 1,3-PDT for 5 h; (b) 3.0 mM 1-PT for 30 min, 3.0 mM 1,3-PDT for 5 h, and hydroquinone (H_2Q) moiety; (c) 2-P-1-T for 30 min, 3.0 mM 1,3-PDT for 5 h and hydroquinone (H_2Q) moiety. The scan potential was from -0.35 V to +0.35 V, at 100 mV/s. In figure 4(a) redox couple was not observed, indicating that the thiol (-SH) group on gold surface is not active at this potential window.

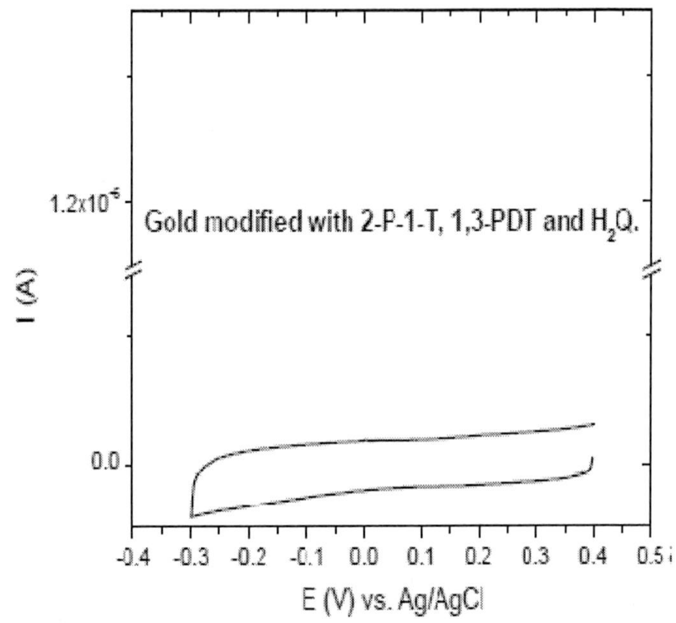

Figure 5. Cyclic voltammograms for the gold modified with 3.0 mM 2-P-1-T for 30 min and 3.0 mM 1,3-PDT for 5 h. The solution was degassed with N_2 flow for least 30 min. Scan rate 100 mV s^{-1}. Measurements were performed at 21 $\pm$ 1 °C.

In Figures 4(b) and 4(c), both have the same behavior en la oxidation and reduction of H_2Q on gold surface modified (H_2Q Q^{2-} + H^+). However, in the Figure 4c was increased the anodic and cathodic peak currents, possibly due to double bond (C=C) in 2-P-1-T. The only difference between these two systems is changing 2-P-1-T by 1-PT, maintaining the same 1.3-PDT and H_2Q on the surface. A cathodic peak potential (E_{pc}) of +0.210 V, an anodic peak potential (E_{pa}) of +0.240 V, and a peak separation (ΔE_p) of 0.030 V, were observed. This Peak-to-peak separation indicates that the overall process is kinetically fast for two-electron redox potential; however, an anodic peak current (i_{pa}) of

0.22 μA and a cathodic peak current (i_{pc}) of -0.10 μA, for a $|i_{pa} / i_{pc}|$ of 2.2, shows a quasi-reversible system. There are substantial differences between the electrochemistry of BQ on clean gold surface and H_2Q moiety on gold surface modified. This system has lower reversibility than the BQ on clean gold surface.

To confirm that the Michael addition reaction between the BQ and the thiol (-SH) (from 1,3-PDT) was reached, was voltammetric sweeps from +0.40 V to -0.30 V. Thus is excluded the presence of BQ, indicating the formation of H_2Q moiety on the gold surface modified. Figure 5 shows the absence of redox behavior for the BQ molecule (as is observed in Figure 4) because this molecule is reduced to H_2Q on the modified gold surface.

Pre-Treatment and Substrates Modification for Surface Analysis

For X-ray photoelectron spectroscopy (XPS) studies, the sample analyses were performed using the Al Kα (15.0 kV at 350 watts) source of a PHI 5600 ci spectrometer. This instrument has a hemispherical analyzer, a toroidal monochromator, and a multichannel detector. The base pressure in the chamber during analysis was less than 1×10^{-9} Torr.

The binding energy values reported in this study were corrected using the C 1s signal of the atmospheric contaminants (285 eV).

High-resolution spectra were recorded at a take-off angle of 45° and 11.75 eV pass energy, with at least 5 minutes for each element.

Different experiment setups were used for the XPS studies. The conditions for each setup are summarized in results of surface analyses.

All substrates used were new and before their modification they were submitted to a cleaning treatment. MAXTEK® Au substrates were used. The polycrystalline gold electrodes used for the XPS measurements were prepared by polishing a 0.32 cm^2 geometric area electrode with 1.0, 0.3, and 0.05 μm Al_2O_3 paste, and then rinsed copiously with nanopure water. Electrode cleanness was verified by cycling the potential in a 1.0 M H_2SO_4 solution, at 100 mV/s scan rate, between hydrogen and oxygen regions until characteristic CV were observed (See Figure 2). The electrode was polished with Al_2O_3 paste, again and washed copiously with nanopure water. Finally, they were dried with an argon gas flow. We verified its cleanness by cyclic voltammetry before being chemically modified.

The previously cleaned substrates were submerged in a 3.0 mM 2-P-1-T ethanol solution for 30 min. afterward, the substrates were washed with ethanol, dried under nitrogen gas flow and submerged in a 3.0 mM 1, 3-PDT ethanol solution for 5 h. Finally, the modified electrodes were immersed in a 2.01 mM 1,4-benzoquinone (BQ) solution, in Britton-Robinson buffer for 15 h. In this last solution, the BQ compound react through a nucleophilic 1,4-Michael addition to form the corresponding reduced thiol conjugate on gold surface.

Results of Spectroscopic Studies

XPS has been widely used to study the composition of self-assembled monolayers on metallic surfaces. This technique was used to confirm the 2-P-1-T, 1,3-PDT, and H_2Q organization on the gold surface. The XPS spectrum of the clean gold substrates is shown in Figure 6. This spectrum showed photoemission peaks that were attributed to the Au and C atoms that are present on the clean gold surface.

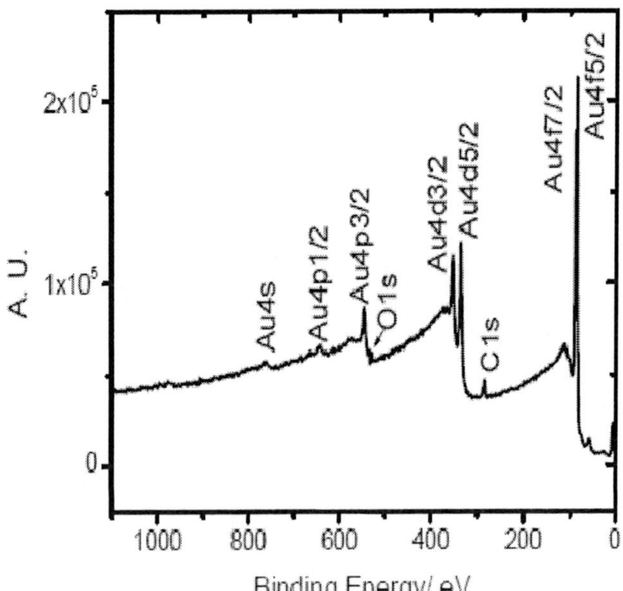

Figure 6. Survey of X-ray photoelectron spectroscopy (XPS) for clean Au surface.

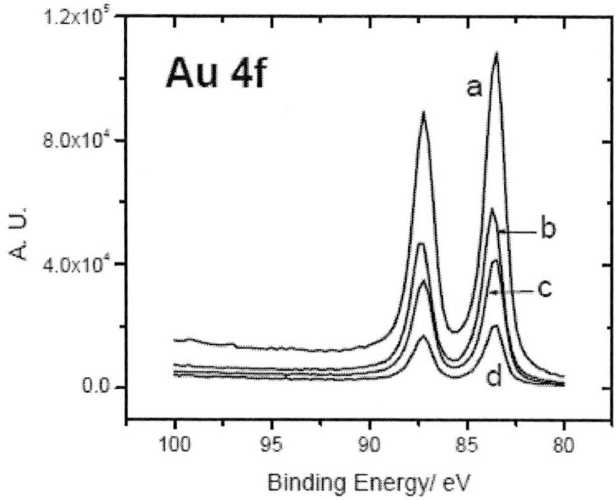

Figure 7. (a) Clean Gold; (b) 3.0 mM 2-P-1-T for 30 min; (c) 3.0 mM 2-P-1-T for 30 min and 3.0 mM 1,3-PDT for 5 h (d) 2-P-1-T for 30 min, 3.0 mM 1,3-PDT for 5 h and hydroquinone (H_2Q) moiety.

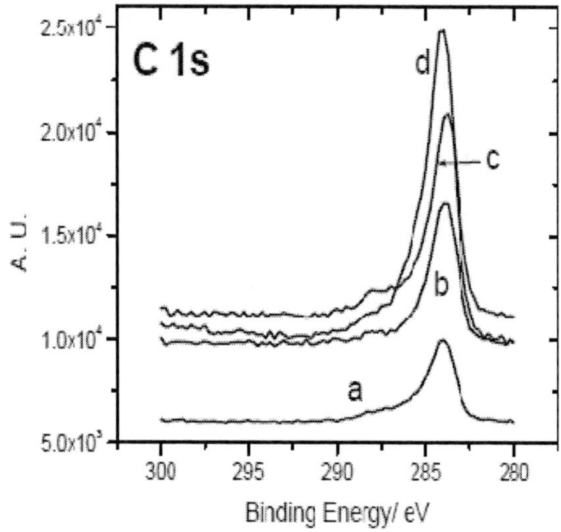

Figure 8. XPS spectra for the O (1s) region **(a)** Clean Gold; **(b)** 3.0 mM 2-P-1-T for 30 min; **(c)** 3.0 mM 2-P-1-T for 30 min and 3.0 mM 1,3-PDT for 5 h **(d)** 2-P-1-T for 30 min, 3.0 mM 1,3-PDT for 5 h and hydroquinone (H_2Q) moiety. The study made at a 45^0 take-off angle.

High-resolution XPS spectra for the core binding energy region of gold are shown in Figure 7. The spectra, taken under the same instrumental conditions, reveal how the signal of gold is reduced when the Au surface was modified with 2-P-1-T, 1,3-PDT, and H_2Q. This indicates that in some way these molecules are adsorbed on the surface and are blocking the XPS detection of Au 4f peaks. In the figure 8 the presence of carbon on the gold surface is observed. The clean gold surface does not present this atom; the signal of carbon from atmospheric contamination only.

Figure 9 shows XPS spectra for the O (1s) region at take-off angles of 45^0. Reveal how the Oxygen signal is increased when the Au surface was modified with 2-P-1-T, 1,3-PDT, and H_2Q, respectively. The O (1s) peak (Figure 9a) appear at a binding energy in accordance with oxides and to a H_2O signal [111,112]. Changes in the area of each O (1s) signal were observed (Figure 9b and 9c), indicating that the concentration of the different types of compound without oxygen is remarkable (2-P-1-T and 1,3-PDT compounds). After the reaction of the 1,3-PDT and BQ compounds, increases the presence of oxygen, indicating that the H_2Q has been formed on the modified surface.

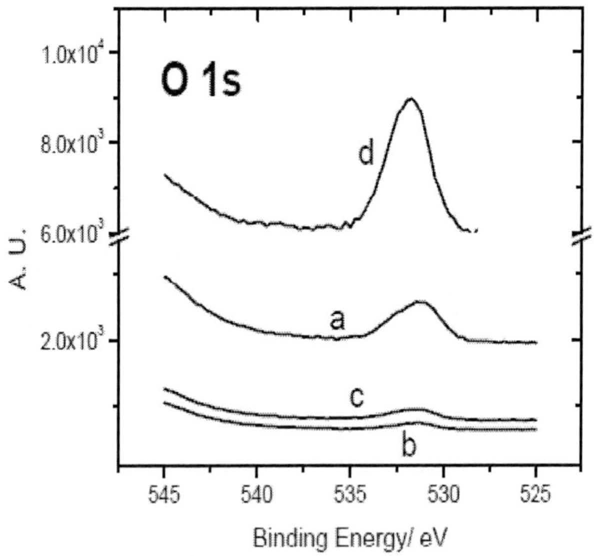

Figure 9. XPS spectra for the O (1s) region **(a)** Clean Gold; **(b)** 3.0 mM 2-P-1-T for 30 min; **(c)** 3.0 mM 2-P-1-T for 30 min and 3.0 mM 1,3-PDT for 5 h **(d)** 2-P-1-T for 30 min, 3.0 mM 1,3-PDT for 5 h and hydroquinone (H_2Q) moiety. The study made at a 45^0 take-off angle.

CONCLUSION

The use of modified metal surfaces is very useful in the study of biomarkers due to their ability and stability in performance.

Owing to the important physiological roles of quinones, such as hydroquinone, 1,4 benzoquinone, and reduced thiols such as GSH, it is necessary to develop sensitive and selective methods for the measurement of those substances to study cellular mechanisms of chemical-induced toxicity. For this purpose, the benzoquinone (BQ) is used in the electrochemical determination of thiol molecules. The Addition of sulfur nucleophiles to quinones has been studied and characterized as 1,4-reductive addition of the Michael type. Quinones, containing a polarized double bond, are considered Michael acceptors, where the Michael donors are thiol nucleophiles. The final product of this reaction is hydroquinone with a sulfur atom in the ortho position. If a metal surface is modified with a dithiol (to form a thiolate with the metal surface, leaving a free thiol,-HS) molecule, the reaction with BQ will be easy to produce hydroquinone (H2Q) moiety. The latter product can be used to determine electrochemically thiol-containing compounds in its structure.

REFERENCES

[1] Nelson D. L, Cox M. M. 2000. *Lehninger Principles of Biochemistry,* third ed. Worth Publishers, New York, NY, p. 842.

[2] Ahmad I. M., Britigan B. E., Abdalla M. Y., *J. Toxicol. Environ. Health* 2011; 74:43–51.

[3] Tepe, N., Shimko L.A, Duran M., *Bioresource Technol.* 2006; 97: 592–598.

[4] Kuśmierek K, Chwatko G, Głowacki R, Kubalczyk P, Bald E, *J. Chromatogr.* B, 2011; 879 (17-18):1290-1307.

[5] Wroński M., *J. Chromatogr. B* 1996; 676:29-34.

[6] Kuśmierek K, Bald E., *Chromatographia* 2008; 67:23-29.

[7] Badaloo A, Reid M, Forrester T, Heird WC, Jahoor F., *Am. J. Clin. Nutr.* 2002;76:646–52.

[8] Meister A, *Methods Enzymol.* 1995; 251: 3–7.

[9] Circu M, Yee Aw T., *Free Rad Res.* 2008; 42(8): 689-706.

[10] Pastore A, Federici G, Bertini E, Piemonte F., *Clin. Chim. Acta* 2003; 333:19-39.

[11] Cross CE, Halliwell B, Borih ET, Proyer WA, Ames BN, Saul RL, McCord JM, Harman D., *Ann. Intern. Med.* 1987; 107:526-545.

[12] Dalle-Donne I, Milzani A, Galgliano N, Colombo R, Giustarini D, Rossi R., *Antioxid. Redox Signal* 2008; 10:445-473.

[13] Ballatori N, Krance SM, Notenboorn S, Shi S, Tieu K., Hammond CL., *Biol. Chem.* 2009; 390:191–214.

[14] Giustarini D, Dalle-Donne I, Tsikas D, Rossi R., *Crit. Rev. Clin. Lab. Sci.* 2009; 46:241-281.

[15] Becker K, Leichsenring M, Gana L, Bremer HJ, Schirmer RH., Free *Radic. Biol. Med.* 1995; 18 :257-63.

[16] Jackson AA., *Trans R Soc. Trop. Med. Hyg.* 1986; 80: 911–3.

[17] Pastore A, Massoud R, Motti C, Lo Russo A, Fucci G, Cortese C, Federici G., *Clin. Chem.* 1998; 44:825-832.

[18] Kosower NS, Kosower EM., *Int. Rev. Cytol.* 1978; 54: 109-160.

[19] Martinez-Huitle C A, Carlesi Jara C, Cerro-Lopez M, Quiroz MA., *Lat. Am. Appl. Res.* 2010; 40(1):47-51.

[20] Han Q, Xu M, Tang L, Sun X, Zhang N, Tan X, Tan X, Tan Y, Hoffman RM., *Clin. Chem.* 2004; 50:1229- 1231.

[21] Kumar A, John L, Alam M, Gupta A, Sharma G, Pillai B, Sengupta S., *Biochem. J.* 2006; 396(Pt 1): 61–69.

[22] Wronska-Nofer T, Norfer JR, Stetkiewicz J, Wierzbicka M, Bolinska H, Fobker M, Schulte H, Assmann G, Eckardstein A., *Nutr. Metab. Cardiovasc. Dis.* 2007; 17: 546–553.

[23] Janaky R, Varga V, Hermaan A, Saransaari P, Oja SS., *Neurochem. Res.* 2000; 25:1397-1405.

[24] Shahrokhian S., *Anal. Chem.* 2001; 73:5972-5978.

[25] Hortin GL, Sullivan P, Csako G., *Clin. Chem.* 2001; 47:1121-1124.

[26] Seshadri S, Beiser A, Selhub J, Jacques PF, Rosenberg IH, D'Agostino RB, Wilson PW, Wolf PA. , *N. Engl. J. Med.* 2002; 346:476-83.

[27] Steegers-Theunissen R, Boers GHJ, Trijbels FJ, Finkelstein JD, Blom HJ, Thomas CM, Borm GF, Wouters MG, Tom Eskes T., *Metabolism* 1994; 43(12):1475-1480.

[28] van Meurs JBJ, Dhonukshe-Rutten RAM, Pluijm SMF, van der Klift M, de Jonge R, Lindemans J,. de Groot L, Hofman A, Witteman J, van Leeuwen J, Breteler M, Lips P, Pols H, Uitterlinden AG., *N. Eng. J. Med.* 2004; 350:2033-2041.

[29] Wu G, Fang,YZ, Yang S, Lupton JR, Turner ND. , *J. Nutr.* 2004; 134:489-492.

[30] Finkelstein JD, Martin JJ. Homocysteine. *Int. J. Biochem. Cell Biol.* 2000; 32:385.

[31] Boushey CJ, Beresford SAA, Omenn GS, Motulsky AG., *J. Am. Med. Asso.* 1995; 274:1049-1057.

[32] Pahor M, Elam MB, Garrison JR, Kritchevsky SB, Applegate WB., *Arch. Intern. Med.* 1999; 159:237-245.

[33] Hogg N., *Free Rad. Biol. Med.* 1999; 27: 28-33.

[34] Etienne S, Matt M, Oster T, Samadi M, Beley M., *Tetrahedron.* 2008; 64:9619–9624.

[35] Hignett G, Threlfell S, Wain AJ, Lawrence N, Wilkins SJ, Davis J, Compton RG, Cardosi M., *Analyst* 2001; 126:353-357.

[36] Tirelli A, Fracassetti D, De Noni I., *J. Agric. Food Chem.* 2010; 58:4565-4570.

[37] Chrastil J., *Analyst* 1990; 115: 1383-1384.

[38] Zaia DA, Ribas KC and Zaia CT., *Talanta* 1999; 50:1003-1010.

[39] Shimada K, Mitamura K., *J. Chromatogr.* 1994; 659:227-241.

[40] White PC, Lawrence NS, Davis J, Compton RG., *Electroanalysis* 2002;14:89-98.

[41] Lawrence N, Deo R, Wang J., Talanta. 2004; 63:443-449.

[42] Lawrence NS, Beckett EL, Davis J, Compton RG., *Anal. Biochem.* 2002; 303:1-16.

[43] Gracheva S, Digga A, Livingston C, Davis J., *Electroanalysis* 2005; 17:205-209.

[44] Finley K. T., In: Patai S., ed. *The chemistry of quinoind compounds.* London; John Wiley and Sons (1974) 877.

[45] Brunmark A., Cadenas E., *Free Radical Biology and Medicine* 7 (1989) 435.

[46] Williams R. J. P., Phil. *Trans. R. Soc. Lond.* B 311:593 (1985) 603.

[47] Swallow A. J., London, Academic Press (1982) 161.

[48] Wardmen P., Clarke E. D., *Br. J. Cancer* 55 (1987) 172.

[49] Hayon E., Simic M., *Acc. Chem. Res.* 7 (1974) 114.

[50] Steenken S., *J. Phys. Chem.* 83 (1979) 595.

[51] Digga A, Gracheva S, Livingstone C, Davis J., *Electrochem. Commun.* 2003; 5:732–736.

[52] Zaborska W, Krajewska B, Kot M, Karcz W., *Bioorg. Chem.* 2007; 35: 233-242.

[53] Bulgariu L, Bulgariu D., *Chem. Bull. "POLITEHNICA" Univ Timisoara* 2008; 53:1-2.
[54] Chambers J., Q. *Chem. Quinonoid. Compd.* Pt. 2 (1974) 737.
[55] Chambers J., Q. *Chem. Quinonoid. Compd.* 2 (1988) 719.
[56] Evans D. H., *Encycl. Electrochem. Elem.* 12 (1978) 1.
[57] Garrett R. H., Grisham C. M., *Biochemistry,* 3rd ed.; Thomson Brooks. Cole: Belmont, CA, 2005, 681.
[58] Aoki K., Chen J., Hua Z., *J. Electroanal. Chem.* 610 (2007) 211.
[59] Quan M., Sanchez D., Wasylkiw M. F., Smith D. K., *J. Am. Chem. Soc.* 129(42) (2007) 12847.
[60] Scott M. Rosendahl and Ian *J. Burgess*, 56 (11) 2011, 4361-4368.
[61] Rojas de Astudillo L., Rivera L., Brito-Gómez R., Tremont R. J., *Journal of Electroanalytical Chemistry* 640 (2010) 56–60.
[62] White P. C., Lawrence N. Davis S. J., Compton R. G., *Analitical Chimica Acta* 447 (2001) 1.
[63] Digga A., Gracheva S., Livingstone C., Davis J., *Electrochemistry Communications* 5 (2003) 732.
[64] Stone C. G., Cardosi M. Davis F. J., *Analytica Chimica Acta* 491 (2003) 203.
[65] Rohner T. Rossier J. S., Girault H. H., *Electrochemistry Communications* 4 (2002) 695.
[66] Wen-Wu Li, Heinze J., Haehnel W., *J. Am. Chem. Soc.* 127(17) (2005) 6140.
[67] Shamsipur M., Kazemi S. H.,Alizadeh A., Mouavi M. F., Workentin M. S., *J. Electroanal. Chem.* 610 (2007) 218.
[68] Ulman A., *Chem. Rev.*, 96 (1996) 1533.
[69] Swalen J. D., Allara D. L., Andrade J. D., Chandross E. A., Garoff S., Israelachvili J., McCarthy T. J., Murria R., Pease R. F., Rabolt J. F., Wyanne K. J., Yu H., *Langmuir.,* 3 (1987) 932.
[70] Laibinis P. E., Whitesides G. W., Allara D. L., Tao Y. T., Parikh Y. T., Unzo R. G., *J. Am. Chem Soc.* 113 (1991) 7152.
[71] Bryant M. A., Pemberton J. E., *J. Am. Chem. Soc.* 113 (1991) 8284.
[72] Bryant M. A., Pemberton J. E., *J. Am. Chem. Soc.* 113 (1991) 3629.
[73] Bryant M. A., Joa S. L., Pemberton J. E., *Langmuir.* 8 (1992) 753.
[74] Sheen C. W., Shi J. X., Martensson J. , Parikh A. N., Allara D. L., *Chem. Phys. Lett.* 114 (1992) 1514.
[75] D. Allara L., Unzo G. R., *Langmuir.* 1 (1985) 45.
[76] Moskovits M., Suh J. S., *J. Am. Chem. Soc.* 107 (1985) 6826.

[77] Schlotter N. E., Porter M. D., Bright T. B., Allara D. L., *Chem. Phys. Lett.* 132 (1986) 93.
[78] Cohen S. R., Naaman R., Savig J., *J. Phys. Chem.* 90 (1986) 3054.
[79] Wasserman S. R., Tao Y. T., Whitesides G. M., *Langmuir.* 5 (1989) 1074.
[80] Finklea H. O., Robinson L. R., Blackburn A., Richter B., Allara D. L., Bright T., *Langmuir.* 2 (1986) 239.
[81] Finklea H. O., Hanshew D. D., *J. Am. Chem. Soc.* 114 (1992) 3173.
[82] Finklea H. O., Ravenscroft M. S., Snider D. A., *Langmuir.* 9 (1993) 223.
[83] Chidsey C. E. D., Loiacono D. N., *Langmuir.* 6 (1990) 682.
[84] Becka A. M., Miller C. J., *J. Phys. Chem.* 96 (1992) 2657.
[85] Collinson M., Bowden E. F., Tarlov M. J., *Langmuir.* 8 (1992) 1247.
[86] Finklea H. O., Hanshew D. D., *J. Electroanal. Chem.* 347 (1993) 327.
[87] Terrettaz S., Becka A. M., Traub M. J., Fettinger J. C., Miller, C. J. *J. Phys. Chem.* 99 (1995) 11216.
[88] Díaz T. L, Malavé M., Rivera M., Castillo J., Cabrera C. R., Rosa Brito, Rolando J. Tremont, 2008, *Applied Surface Science.* 254 (6), 1587-1592.
[89] Frei M., Diederich F., Tremont R. J., Díaz T. L., Echegoyen L., 2006, *Helvetica Chimica Acta,* 89 (9) 2040–2057.
[90] Morales-Cruz, A.,L, Tremont, R.J., Martínez, R., Romañach, R., and Cabrera, C.R.. 2005, *Applied Surface Science.* 241(3-4): 371-383.
[91] Brito, R., Tremont, R.J., and Cabrera, C.R. 2004, *J. Electroanalytical Chemistry.* 574: 15-22.
[92] Tremont, R.J., Cruz, G., and Cabrera, C.R. 2003, *J. Electroanalytical Chemistry.* 558: 65-77.
[93] Tremont, R.J., Blasini, D., and Cabrera, C.R. 2003, *Journal Electroanalytical Chemistry.* 556:147-158.
[94] Brito, R, Tremont R.J., Feliciano, O., and Cabrera, C.R. 2003, *J. Electroanalytical Chemistry.* 540: 53-62.
[95] Cabrera, C.R., Warner, J.D., Mueller. C.H., Van Keuls, F., Miranda, F.A., Tremont, R.J., Blasini, D., and Morales, A.. 2001, *Materials Research Society (MRS) Proceedings.* 666, F9.1/1-F9.1/6.
[96] Blasini D.R, Tremont R.J., and Cabrera C.R, Batina, N., and González, I. 2003, *J. Electroanalytical Chemistry.* 540: 45-52.
[97] Tremont R.J., and Cabrera C.R. 2002, *J. of Applied Electrochemistry.* 32: 783-795.
[98] Xu J., Li H. L., Zhang Y. J., *J. Phys. Chem.* 97 (1993) 11497.
[99] Rubinstein I., Steinberg S., Tor Y., Shanzer A., Sagiv J., *Nature* 332 (1998) 426.

[100] Steinberg S., Tor Y., Sabatini E., Rubinstein I., *J. Am. Chem. Soc.* 113 (1991) 5176.

[101] Steinberg S., Rubinstein I., *Langmuir.* 8 (1992) 1183.

[102] Finklea H. O., Snider D. A., Fedyk J., *Langmuir. 6* (1990) 371.

[103] Finklea H. O., Snider D. A., Fedyk J., Sabatini E., Gafni Y., Rubinstein I., *Langmuir* 9 (1993) 3660.

[104] Miller C. J., Cuendet P., Grätzel M., *J. Phys. Chem.* 95 (1991) 877.

[105] Miller C. J., Grätzel M., *J. Phys. Chem.* 95 (1991) 5225.

[106] Laibinis P. E., Whitesides G. W., Allara D. L., Tao Y. T., Parikh Y. T., Unzo R. G., *J. Am. Chem Soc.* 113 (1991) 7152.

[107] Yang W., Gooding J. J., Hibbert D. B., *J. Electroanal. Chem.* 516 (2001) 10.

[108] Becka A. M., Miller C. J., *J. Phys. Chem.* 96 (1992) 2657.

[109] Miller C. J., Cuendet P., Grätzel M., *J. Phys. Chem.* 95 (1991) 877.

[110] Stone C. G., Cardosi M. Davis F. J., *Analytica Chimica Acta* 491 (2003) 203.

[111] Moulder J. F., Stickle W. F.. Sobol P. E., K. D. Bomben In Handbook of X-ray Photoelectron Spectroscopy; Chastain, J., Ed.; Perkin Elmer Corporation, Physical Electronics: Eden Prelectrolyteie, M.N., 1992.

[112] Laibinis P. E., Nuzzo R. G., Whitesides G. M., *J. Phys. Chem.* 1992, 96, 5097.

In: Hydroquinone
Eds: F. Gokden and A. Lazzarotto
ISBN: 978-1-62100-258-1

Chapter 5

CELLULAR EFFECTS OF HYDROQUINONE

T. B. Ng, X. J. Ye and J. H. Wong
School of Biomedical Sciences,Faculty of Medicine,
The Chinese University of Hong Kong, Shatin,
New Territories,Hong Kong, China

ABSTRACT

Hydroquinone (HQ) is a major benzene metabolite that can be further oxidized toreactive quinones. It has been employed as a component of skin lighters and a photographic reducer.Its adverse effects on different cells have been demonstrated.High concentrations of HQ induces apoptosis in Jurkat cells.

It induces DNA and chromosomal damage in mouse fibroblasts. It elicits DNA hypomethylation that accounts for the leukomogenicity of benzene.HQ causes apoptosis in human peripheral white blood cells.

It induces oxidative damage anddiminishes expression of human 8-oxo-guanine DNA glycosylase mRNA in human A549 lung adenocarcinoma cells. It hinders progression of T cell cycle by suppressingcyclin-dependent kinases.It brigs about alterations in protein expression inhuman hepatocytes. Exposure to HQ in utero causes a decline in the percentage ofdifferentiated HD3 avian erythroblast cells concomitant with a rise in reactiveoxygen species, both of which can be reduced by superoxide dismutase conjugatedto polyethylene glycol. Myeloperoxidase-activated HQ inhibits topoisomerase II,a.

INTRODUCTION

Hydroquinone, a major metabolite of benzene has been employed as an antioxidant, a stabilizer, a photographic reducer, and an ingredient in skin lighteners. It is used in cosmetics because of its fragrance, antioxidant, reducing, or polymerization inhibitor activities. Since January 1, 2001, its use in cosmetics has been banned. This ban is due to mid-term effects such as leukoderma-en-confetti/occupational vitiligo and exogenous ochronosis. However, a recent literature search on hydroquinone as a skin lightening agent suggests that possible long-term effects such as carcinogenesis may be expected as well. Metabolites of hydroquinone formed in the liver, e.g., p-benzoquinone and glutathione conjugates of hydroquinone, are mainly responsible for this. In the bone marrow, hydroquinone is oxidized into p-benzoquinone because of the high myeloperoxidase activity. Topically applied hydroquinone-containing creams may give rise to accumulation of these compounds, which can cause DNA damage and mutations. They also have the capability to disrupt protective mechanisms, whereby they facilitate further development of cancer. In the bone marrow, long-term effects such as aplastic anemia and acute myeloid leukemias may occur. Most of the evidence stems from research on benzene toxicity, which appears to arise via its metabolite hydroquinone. There is no report yet demonstrating carcinogenesis resulting from the application of hydroquinone-containing creams. However, doctors should be aware of these potential health risks which were up until now disregarded. (Westerhof and Kooyers, 2005).

The following oi a review of the effects of hydroquinone with an emphasis on its actions on different cell types.

EFFECT ON PIGMENTATION

Postinflammatory hyperpigmentation (PIH) is reactive hypermelanosis and consequence of a diversity of inflammatory skin conditions. It can exert an adverse effect on a patient's quality of life, especially for patients with a darker complexion. Dyschromias, including PIH, are one of the most common presenting complaints of racial ethnic groups with a darker skin color. This is attributed to an augmented production or deposit of melanin in the epidermis or dermis by labile melanocytes. Different exogenous or endogenous inflammatory conditions can lead toPIH and typically most epidermal lesions appear brown while dermal hypermelanosis appears blue-gray. Depigmenting agents target different steps in melanogenesis. Most of them, hydroquinone being one example, inhibit tyrosinase. (Callender et al., 2011).

Skin depigmentation is a therapeutic measure for extensive vitiligo and may be useful for forestalling tumor recurrences and as a prophylactic treatment of familial melanoma. Prevention of melanoma necessitates a bleaching agent-induced loss of exposed melanocytes supported by an immune response to distant pigment cells. Investigations on human explant cultures treated with depigmenting agents like 4-tertiary butyl phenol (4-TBP) or monobenzyl ether of hydroquinone (MBEH) revealed an enhanced migration of Langerhans cells toward the dermis only following treatment with MBEH, thus indicating selective induction of an immune response. MBEH but not 4-TBP induced significant skin depigmentation in both C57BL/6 wild type as well as k14-SCF transgenic, epidermally pigmented mice. Cytokine expression patterns in skin treated with MBEH support activation of a Th1-mediated immune response corresponding to an influx of T cells and macrophages. Although tumor cells were insensitive of to MBEH-induced cytotoxicity, tumor growth in B16 challenged k14-SCF mice pretreated with MBEH was retarded. This was attributed due to an abundance of cytotoxic T cells accompanied by an increased expression of Th1 and Th17 cytokines. The results suggest that MBEH can be used as a prophylactic treatment for melanoma (Hariharan et al., 2011).

Safety and toxicity studies show that hydroquinone is dermally absorbed in humans from both aqueous and alcoholic formulations and is excreted in the form of glucuronide or sulfate conjugates. An altered immune function in vitro and in vivo in animals and an elevated incidence of renal tubule cell tumors and leukemia in F344 rats due to hydroquinone have been reported, butwhether the findings are relevant to humans is not known. Quantitatively, however, the use of hydroquinone in cosmetics is unlikely to cause renal neoplasia Hydroquinone is safe at concentrations of or below1% in hair dyes and is safe for use in nail adhesives. It should not be used in other leave-on cosmetics (Andersen et al., 2010).

Exogenous ochronosis, an infrequent dermatosis characterized as a dark blue hyperpigmentation, may be caused by the use of antimalarials and substances such as phenol, resorcinol, benzene, or hydroquinone. Ribas et al. (2010) presented four cases of female patients who developed facial hyperpigmentation after the use of compounds containing hydroquinone, characterized clinically and histologically as ochronosis. There remains the possibility of exogenous ochronosis cases misdiagnosed as failure of melasma treatment, and the risk of indiscriminate use of hydroquinone containing compounds without medical prescription (Ribas et al., 2010).

EFFECT ON MELASMA

Melasma is a common acquired symmetric hypermelanosis characterized by irregular light- to gray-brown macules and patches involving sun-exposed areas of skin. Genetic influences, exposure to UV radiation, pregnancy, hormonal therapies, cosmetics, phototoxic drugs, and antiseizure medications are known to play a role in the etiology of melasma. Prevailing treatments for melasma comprise hypopigmenting agents, chemical peels, and lasers. Hypopigmenting agents encompass phenolic and nonphenolic derivatives. Hydroquinone and hydroquinone combination preparations are the phenolic agents employed. Notwithstanding disputes over the issue of hydroquinone-induced ochronosis, hydroquinone is the most effective topically applied bleaching agent approved by the Food and Drug Administration for melasma.treatment .Nonphenolic bleaching agents comoprise tretinoin and azelaic acid. Superficial, medium, and deep chemical peels are usually applied in patients with a lighter-complexion. Caution has to be exercised when it is used in blacks. Although lasers are effective in the therapy of a number of hyperpigmentary disorders, their precise efficacy and place in the therapy of melasma have yet to be corroborated (Grimes,1995).

Costa et al.(2010) conducted a study to assess the clinical efficacy and safety of the association of Belides, Emblica and Licorice 7%, compared to hydroquinone 2%, in the therapy of melasma.

Following two months of exclusive use of an SPF35 sunscreen, 56 women, 18 to 60 years of age, phototypes I to IV, with epidermal or mixed melasma, were allocated into two different groups in a mono-blind clinical study: (group A) cream with Belides, Emblica and Licorice, applied twice a day;(group B) cream with hydroquinone 2%, used at night. They were observed in a 60-day study; every 15 days, they were submitted to medical evaluation, self-evaluation, and photographic registration (Visia(®)).

There were no statistically significant differences between groups in the improvement of melasma. Group A exhibited less adverse dermatological reacrtions. Therefore, the association of Emblica, Licorice and Belides is a safe and efficient alternative for the therapy of melasma.

EFFECT ON HUMAN BLOOD CELLS

Exposure to benzene leads to reduction of peripheral blood cell counts, aplastic anemia and leukemia. Zhao et al.(2009) analyzed differential gene expression in peripheral white blood cells from seven female patients who were victims of occupational benzene poisoning and seven matched control subjects.

In all patients, enhanced expression of CYP4F3A encoding the leukotriene B4 (LTB(4)) omega-hydroxylase important for the inactivation of LTB(4) in polymorphonuclear leukocytes was observed. Four patients exhibited augmented expression of CYP1A1 while augmented expression of CYP1B1 was noted in two patients. Changes in expression of CYP2B6, CYP51, and CYP27A1 were detected in some of the patients. Mechanistic studies disclosed that phenol, a major metabolite of benzene, upregulated the expression of CYP4F3A at both mRNA and protein levels in cultured promyelocytic leukemia cells (HL-60), in differentiated HL-60 cells, and in the proerythroid cell line K562, and ex vivo in human neutrophils.However, hydroquinone brought about massive apoptosis of the cells. The results show that benzene and metabolites induce CYP4F3 in human blood cells both in vivo and in vitro and that induction of CYP4F3 may have a role in the development of benzene hematotoxicity and serve as a biomarker of benzene exposure.

EFFECT ON CHICKEN ERYTHROBLAST HD3 CELLS

Benzene is a known leukemogen that is metabolized to form reactive intermediates and reactive oxygen species (ROS). c-Myb oncoprotein is a transcription factor witha crucial role in hematopoiesis. and c-Myb transcript and protein are overexpressed in a number of leukemias and cancers. In view of the role of c-Myb' in hematopoiesis and leukemias, it is likely that benzene interferes with the c-Myb signaling pathway and that this involves ROS. Wan and Winn (2007) observed that in chicken erythroblast HD3 cells, benzoquinone, hydroquinone, and catechol heightened DCFDA fluorescence, enhanced dihydrorhodamine-123 (DHR-123) fluorescence, diminished GSH: GSSG ratios, and upregulated reporter construct expression after 24 h of exposure. SOD could block 5-(and-6)-chloromethyl-2',7'-dichloro dihydrofluorescein diacetate (DCFDA) fluorescence and c-Myb activity induced by benzoquinone and hydroquinone only.

These data indicate metabolite differences in benzene-mediated toxicity. This investigation supports the hypothesis that benzene may mediate its toxicity through ROS-mediated alterations in the c-Myb signaling pathway. (Wan and Winn, 2007).

EFFECT ON HEMATOPOIETIC PROGENITOR CELLS

Hydroquinone (HQ) modifies cytokine response and induces cell death in hematopoietic progenitor cells (HPC), and inhibits nuclear factor kappa B (NF-kappaB) activation in T and B cells. In both hematopoietic cell line (TF-1) and primary HPC TNF-alpha activated NF-kappaB. HQ exposure suppressed activation of NF-kappaB by TNF-alpha. HQ sensitized the cells to the pro-apoptotic effect of TNF-alpha. NF-kappaB plays an important role in HPC survival, and HQ-induced inhibition of NF-kappaB makes these cells susceptible to cytokine-induced apoptosis. These effects may play a role in the inhibition of hematopoiesis in individuals exposed to benzene (Kerzic et al., 2003).

EFFECT ON ERYTHRID PROGENITOR-LIKE CELLS

Benzene is a common occupational hazard and a ubiquitous environmental pollutant. The metabolites of benzene play important roles in the toxicity of benzene to the hematopoietic system, but little is known about the effects of benzene metabolites on erythropoiesis. In the study of Wu et al. (2011), erythroid progenitor-like K562 cells were used to determine the effects of phenolic metabolites of benzene, including phenol, hydroquinone and 1,2,4-benzenetriol, on the erythroid differentiation. After the treatment with these benzene metabolites at concentrations that one not cytotoxic, the hemin-induced hemoglobin synthesis in K562 cells underment decline in a concentration- and time-dependent manner, Expression of CD71 and GPA protein on the surface of K562 cells was also reduced. The reverse transcription-PCR was used to determine the mRNA level of the erythroid related genes in the K562 cells that were treated with benzene metabolites. The hemin-induced expression of globin genes, including α-, β- and γ-globin genes, and the gene encoding the heme synthesis enzyme porphobilinogen deaminase was suppressed by benzene metabolites. When the K562 cells were pretreated with benzene metabolites, the hemin-induced expression of two transcription factor genes GATA-1 and NF-E2 diminished, and the pre-treatment with benzene metabolites downregulated the mRNA level of transcription factor gene GATA-2 by hemin.

These results indicated that benzene metabolites inhibited the hemin-induced erythroid differentiation through affecting the transcription of the erythroid related genes. (Wu et al., 2011)

EFFECT ON HUMAN BONE MARROW MONONUCLEAR CELLS

HQ curtailed the protein and mRNA expression of topoisomerase IIα(TOPOIIα) in human bone marrow mononuclear cells. The change of histone chemical modification (histone H3 acetylation) is important to the hematopoietic toxicity of benzene Shi et al., 2010).

EFFECT ON HUMAN EOSINOPHILS AND NEUTROPHILS

Hydroquinone (HQ) is involved in hematopoiesis due to its accumulation in bone marrow. It affects various immune responses. HQ markedly increased the apoptosis of neutrophils and eosinophils which function as important leukocytes in immunological regulation and immune diseases. In this study, we examined the toxic effects of HQ on the apoptosis of human neutrophils and eosinophils isolated from the blood of healthy donors. The pro-apoptotic effect was involved in activation of caspase 9 and caspase 3. Reactive oxygen species (ROS) production was enhanced after HQ treatment in a dose-dependent manner.

The release of IL-8 and MCP-1 from neutrophils and eosinophils, respectively, was potentiated.The results of this study showed that HQ strongly induces the apoptosis of neutrophils and eosinophils through the caspase 9/3-dependent pathway and the enhanced ROS formation. HQis cytotoxic to human neutrophils and eosinophils and may adversely affect the regulation of immune responses (Yang et al., 2011).

EFFECT ON HUMAN BASOPHILS

Benzene and its metabolites are implicated in the pathogenesis of chronic lung inflammation and allergic diseases including bronchial asthma.To investigate the actions of these xenobiotics on human basophilswhich play a crucial role in the development of respiratory allergy ,Borrelli et al.(2010) studied the actions of hydroquinone (HQ) and benzoquinone (BQ), two important chemicals implicated in benzene toxicity, on the release of preformed (histamine) and de novo synthesized mediators (cysteinyl leukotriene C4, LTC4, and IL-4) from human basophils. Histamine release from basophils from normal donors induced by anti-IgE was inhibited up to 30% by HQ (3-100 microM) and that induced by the Ca2+ ionophore A23187 up to 55%. HQ had no effect on histamine release induced by formyl-methionyl-leucyl-phenylalanine (f-Met-Leu-Phe). BQ was slightly more potent than HQ.HQ completely suppressed the de novo synthesis of LTC4 from basophils challenged with anti-IgE or f-Met-Leu-Phe and the production of IL-4 in cells stimulated with anti-IgE. Thus HQ and BQ inhibit the release of proinflammatory mediators and Th2-promoting cytokines from basophils activated by different stimuli.

EFFECT ON JURKAT CELLS AND HL60 CELLS

The actons of hydroquinone on Jurkat cells have been studied by Kim et al.(2009).Intracellular reactive oxygen species (ROS) levels were elevated. There was a substantially enhanced phosphorylation of ERK and activation of caspase-9/-3 accompanied by PARP cleavage. HQ-induced apoptosis was abrogated by the ERK inhibitor PD98059 or N-acetylcysteine (NAC) . ERK protein, but not JNK or p38, was activated. However, S-phase recruitment was reduced following preincubation with NAC, though not by PD98059.. Hence it appears that elevated levels of ROS brought about HQ-induced apoptosis via ERK signaling and the caspase pathway. On the other hand, low levels of ROS elicited S-phase recruitment in the cell-cycle distribution (Kim et al., 2009).

Hydroquinone produces its myelotoxicity by conversion to reactive quinone derivatives in myeloperoxidase (MPO)-containing cells. Inayat-Hussain et al. (2001) conducted an investigation to examine the role of caspases in hydroquinone-induced apoptosis in MPO-rich HL-60 promyelocytic leukemia and MPO-deficient Jurkat T-lymphoblastic leukemia cells Apoptosis was followed by phosphatidylserine (PS) exposure, caspases-3/-7 activation, PARP cleavage, DNA fragmentation, and ultrastructural changes.. In HL-60 cells, the general caspase inhibitor benzyloxycarbonyl-Val-Ala-Asp fluoromethyl ketone (Z-VAD.FMK) prevented activation of caspases-3/-7, cleavage of PARP, and DNA, but PS externalization and cytoplasmic changes were unaffected. To the contrary, all apoptotic features were abolished by Z-VAD.FMK in HQ-treated Jurkat cells. These data provide evidence for Z-VAD.FMK-insensitive and caspases-3/-7-independent pathway(s) in the externalization of PS and cytoplasmic changes during HQ-induced apoptosis in HL-60 cells.However, in Jurkat cells, all of the aforementioned changes were dependent on caspase activation. The fact that MPO-deficient Jurkat cells as well as MPO-rich HL-60 cells undergo apoptosis in response to hydroquinone suggests that MPO-catalyzed bioactivation of HQ is not required for toxicity. The disparate mechanisms of apoptosis in HL-60 and Jurkat T cells may reflect the MPO activity of these cells and, the quantity of reactive BQ and other metabolites produced.

Effect on Hl-60 Cells

Zhang et al. (2007) examined the effect of TTRAP expression on hydroquinone -induced apoptosis in HL-60 cells in vitro, and investigated the relationship between TTRAP expression and apoptosis.The apoptotic rate of HL-60 cells and the amount of TTRAP expression in the mRNA level increased as the concentrations of hydroquinone increased When the concentration of hydroquinone exceeded 250 micromol/L, the rate of necrosis rose abruptly, and TTRAP expression was reduced.

Effect on HeLa Cells

Werner syndrome is a progeroid disorder brought about by WRN gene mutations. The encoded WRN protein is an RecQ helicase that maintains genomic stability. Single nucleotide polymorphisms in WRN are associated with a heightened risk for some cancers and linked to benzene hematotoxicity. To investigate the role of WRN in benzene toxicity, Galvan et al. (2008) used RNA interference (RNAi) to silence WRN in HeLa cells and examined the susceptibility of WRN-depleted HeLa cells to the toxic effects of hydroquinone. HeLa cells were used as the experimental model because RNAi is highly effective in this system producing nearly total elimination of the target protein. WRN depletion ensued in a decrease in cell proliferation and an enhanced susceptibility to hydroquinone cytotoxicity as revealed by an increase in necrosis. After exposure to hydroquinone, WRN-depleted HeLa cells also displayed a rise in the number of DNA double-strand breaks, and an increased DNA damage response. The results indicate that WRN protects HeLa cells against the toxicity of hydroquinone, specifically in eliciting a normal DNA damage response following the induction of DNA double-strand breaks.

EFFECT ON HUMAN LUNG EMBRYONIC FIBROBLASTS

Treatment of MRC-5 human lung embryonic fibroblasts with low levels of hydroquinone ensued in adaptation to further exposure to lethal doses of hydroquinone at the cell-survival level, measured using the alamarBlue assay, lactate dehydrogenase leakage assay and Annexin V-FITC/PI staining., As revealed by two-dimensional electrophoresis combined with mass spectrometry, proteins associated with energy metabolism, protein folding, redox regulation, cell structure and cell signaling were involved in the adaptive response. The results indicate that hydroquinone-induced adaptive response is involved in modulation of diverse cellular functions, and that the redox regulation might be a common mechanism during the adaptive response.(Li et al., 2008).

EFFECT ON MOUSE FIBROBLASTS

The cellular ROS level, DNA damage and micronucleus formation after treatment with HQ were the highest/ most serious in pol beta null cells (pol beta -/-) than pol beta wild-type cells (pol beta +/+), and pol beta overexpressed cells (pol beta oe). The ability of pol beta oe cells to repair DNA damage was more prominent than pol beta +/+ and pol beta -/- cellsThus it appears that cellular ROS generation is induced by HQ, resulting inDNA and chromosomal damage, DNA polymerase beta overexpression could help cells to withstand oxidative damage and protect them from HQ-. induced genotoxicity (Chen et al., 2010).

EFFECT ON HUMAN LIVER L-02 CELLS

After exposure of human liver L-02 cells to HQ, expression of Rho GDP dissociation inhibitor GDI alpha, 6-phosphogluconolactonase, erbB3 binding protein EBP1 and lamin A/C, isoform 1 precursor, which are involved in cell skeleton, signal transduction and energy metabolism, was significantly altered (Ju et al., 2006).

EFFECT ON RAT HEPATOCYTES

Polychlorinated biphenyls (PCBs) are persistent toxic organic pollutants . The mechanisms of toxic action of PCB have been ascribed to toxicological actions of its metabolites, such as hydroquinones, produced by cytochrome-P-450 oxidation. The actions of PCB hydroquinone metabolites on freshly isolated rat hepatocytes were studied by Chan et al. (2010). Hydroquinones can be oxidized to semiquinones and/or quinone metabolites capable ofconjugating glutathione or oxidizing glutathione as a consequence of redox cycling. This lowers the level of glutathione in hepatocytes, an inhibitor of rellular defence mechanisms, producing cell death and an enhanced susceptibility to oxidative stress. However, glutathione-depleted hepatocytes were more resistant to the hydroquinone metabolites of PCBs, indicating that their glutathione conjugates were toxic and that there was a third type of quinone toxicity mechanism entailing a hydrogen peroxide-accelerated autoxidation of the hydroquinones to generate toxic electrophilic quinone and semiquinone-glutathione conjugates.

EFFECT ON HEK293 EMBRYONIC KIDNEY CELLS

HQ induced apoptosis in HEK293 human embryonic kidney cells in a dose-dependent manner. Intracellular thiol, including glutathione and protein thiol, underwent a decline after exposure to HQ. The apoptosis was inhibited by N-acetylcysteine (NAC), a precursor of intracellular GSH synthesis,.and increased by buthionine sulfoximine (BSO), a specific inhibitor of intracellular GSH synthesis. Thus HQ induces apoptosis in HEK293 cells by depleting intracellular thiol. (Shen et al., 2003).

EFFECT ON CHINESE HAMSTER OVARY CELLS

Exposure of humans to environmental benzene results in deleterious chronic effects-even at the genetic level. Pandey et al.(2009) studied in vitro genotoxicity of benzene (BZ) and its major metabolites [p-benzoquinone (BQ), hydroquinone (HQ), catechol (CT), 1,2,4-benzenetriol (BT) and trans-trans muconic acid (MA)] at concentrations 0.5-50 microM, in Chinese hamster ovary (CHO) cells by using a series of assays. The tests included the alkaline Comet assay, cytokinesis blocked micronucleus (CBMN) assay, flow cytometric analysis of micronucleus (flow MN) and chromosome aberration (CA) test. The results disclosed significant concentration-dependent response in all parameters examined. HQ exhibited the highest DNA damaging potency in the Comet assay followed by BQ>BT>CT>BZ>MA. Both CBMN and flow MN assays revealed a good correlation in their results, where BQ and MA exhibited maximum and minimum micronucleus induction respectively. Significant chromosomal abnorrnalities occurred in response to BQ, BT and HQ, with moderate response elicited by CT and BZ and the weakest response to MA. The results showed that Comet assay and flow cytometry are useful for determination of MN, and quantification of in vitro genotoxicity at low levels and also indicated that partially unrepaired DNA damage could exert harmful effects onhealth in people exposed to benzene. Results of in silico studies using different endpoints of genotoxicity and molecular docking studies with human topoisomerase-II alpha, a major DNA repair enzyme, were consistent with the in vitro data, implying a direct relationship of the observed genotoxicity with the structural properties and various interactions of metabolites with the enzyme. Thus the genotoxicity of benzene in mammalian cells is mainlyattributed to inhibition of topoisomerase by the metabolites.

EFFECT ON RETINAL PIGMENT EPITHELIUM

Age-related macular degeneration (AMD) is the leading cause of legal blindness in the elderly population. Debris (termed drusen) below the retinal pigment epithelium (RPE) have been recognized as a risk factor for dry AMD and its progression to wet AMD, which is characterized by choroidal neovascularization (CNV). Pons and Marin-Castano (2011) investigated the expression of pro-inflammatory monocyte chemoattractant protein-1 (MCP-1), pro-angiogenic vascular endothelial growth factor (VEGF) and anti-angiogenic pigment epithelial derived factor (PEDF) in RPE from smoker patients with AMD. The effects of hydroquinone (HQ), a major pro-oxidant in cigarette smoke on MCP-1, VEGF and PEDF expression in cultured ARPE-19 cells andRPE RPE/choroids from C57BL/6 mice were also evaluate. MCP-1 protein was found in RPE from AMD smoker patients at reduced levels. Both MCP-1 mRNA and protein were diminished in ARPE-19 cells and RPE/choroids from C57BL/6 mice after 5 days and 3 weeks of exposure to HQ-induced oxidative injury. VEGF protein expression was unregulated and PEDF protein expression underwent a decline in RPE from smoker patients with AMD versus controls resulting in an increasedVEGF/PEDF ratio. The VEGF/PEDF ratio in vitro and in vivo. Elevated following treatment with HQ for 5 days and 3 weeks, respectively. It was suggested that impaired RPE-derived MCP-1-mediated scavenging macrophages recruitment and phagocytosis might result in incomplete clearance of proinflammatory debris and infiltration of proangiogenic macrophages which together with a higher VEGF/PEDF ratio favoring angiogenesis might promote the accumulation of drusen and progression to CNV in smker patients with dry AMD.

EFFECT ON HUMAN LUNG EPITHELIAL CELLS

Benzene is a well-established human carcinogen. Benzene metabolites hydroquinone (HQ) and benzoquinone (BQ) are highly reactive molecules capable of producing reactive oxygen species and causing oxidative stress. Rubio et al.(2011) examined the role of the Nrf2, a key nuclear transcription factor that regulates antioxidant response element (ARE)-containing genes, in defense against HQ- and BQ-induced cytotoxicity in cultured human lung epithelial cells (Beas-2B). When the cells were exposed to HQ or BQ the activity of an ARE reporter was induced in a dose-dependent manner, meanwhile Nrf2 protein levels were elevated and accumulated in the nucleus. Increased expression of well-known Nrf2-dependent proteins including NQO1, GCLM, GSS and HMOX was also observed in the HQ/BQ-treated cells. Moreover, transient overexpression of Nrf2 conferred protection against HQ- and BQ-induced cell death, whereas knockdown of Nrf2 by small interfering RNA resulted in increased apoptosis. The increased susceptibility of Nrf2-knockdown cells to HQ and BQ was associated with reduced glutathione levels and loss of inducibility of ARE-driven genes, suggesting that deficiency of Nrf2 impairs cellular redox capacity to counteract oxidative damage. The results indicate that Nrf2-ARE pathway is important for protection against HQ- and BQ-induced toxicity.

Effect on Human Bronchial Epithelial Cells

Hu et al., (2007) investigated the toxicological mechanism of hydroquinone in human bronchial epithelial cells and the involvement of DNA polymerase beta in protecting cells from hydroquinone -induced damage. Normal human bronchial epithelial cells and cells transfected with the empty vector of pEGFP-C1 were employed as controls. DNA polymerase beta knock-down cell line was established via RNA interference as an experimental group MTT assay showed that DNA polymerase beta knock-down cells treated with hydroquinone had a lower absorbance value at 490 nm than the control cells. Comet assay revealed that hydroquinone elicited more severe DNA damage in DNA polymerase beta knock-down cell line than in control cells. There was no significant difference in the two control groups. The results suggest that DNA polymerase beta protects cells from hydroquinone- induced damage.

Interaction with DNA

Both synthetic 2-phenyl-1,4-benzoquinones (BQ) and 2-phenyl-1,4-hydroquinones (HQ) derivatives of polychlorinated biphenyls (PCBs) can covalently bind to DNA, and chromatographic similarity in adduct patterns from these two metabolites imply in.

Acute and Subchronic Toxicity

Dermal application of 2000 mg/kg HQ to rabbits under an occlusive wrap for 24 h did not lead to neurobehavioral effects or mortality. Doses of 200 or 64 mg/kg HQ produced acutely observable behavioral effects such as tremors and diminished activity. No morphological lesions of the CNS and PNS, were associated with HQ administration or secondary to repetitive CNS stimulation by HQ. Oral HQ doses of or higher than 64/mg/kg produced acute neurobehavioral effects suggesting CNS stimulation. However, subchronic exposure to dose levels that produced repetitive CNS stimulation by HQ did not ensue in an exacerbation of acute stimulatory effects over time or morphologic changes in the CNS or PNS or nephrotoxicity. (Topping et al., 2007)

Hydroquinone has been used for decades as a skin lightening agent. Since January 1, 2001, its use in cosmetics has been banned. This ban is due to mid-term effects such as leukoderma-en-confetti/occupational vitiligo and exogenous ochronosis. However, a recent literature search on hydroquinone as a skin lightening agent suggests that possible long-term effects such as carcinogenesis may be expected as well. Metabolites of hydroquinone formed in the liver, e.g., p-benzoquinone and glutathione conjugates of hydroquinone, are mainly responsible for this. In the bone marrow, hydroquinone is oxidized into p-benzoquinone because of the high myeloperoxidase activity. Topically applied hydroquinone-containing creams may give rise to accumulation of these compounds, which can cause DNA damage and mutations. They also have the capability to disrupt protective mechanisms, whereby they facilitate further development of cancer. In the bone marrow, long-term effects such as aplastic anemia and acute myeloid leukemias may occur. Most of the evidence stems from research on benzene toxicity, which appears to arise via its metabolite hydroquinone. There is no report yet demonstrating carcinogenesis resulting from the application of hydroquinone-containing creams. However, doctors should be aware of these potential health risks which were up until now disregarded. (Westerhof and Kooyers, 2005).

The potential toxicity of HQ is dependent on the route of exposure, and toxicity in rodents is highly sex-, species-, and strain-specific. Nephrotoxicity in male F-344 rats has been reported. Dermal, developmental and reproductive toxicity studies in rats and rabbits produced negative results. Cancer bioassay data disclose limited effects which are not enough to classify HQ as a human carcinogen. Epidemiological and occupational studies of workers extensively exposed to HQ did not reveal adverse systemic health consequences or carcinogenicity (O'Donoghue, 2006).

Conclusion

From the foregoing account it can be seen that hydroquinone exerts toxicity on a diversity of cells. Extreme caution should be exercised when using any product known to contain this benzene metabolite.

REFERENCES

Andersen FA, Bergfeld WF, Belsito DV, Hill RA, Klaassen CD, Liebler DC, Marks JG Jr, Shank RC, Slaga TJ, Snyder PW. Final amended safety assessment of hydroquinone as used in cosmetics. *Int. J. Toxicol.* 2010;29(6 Suppl):274S-87.

Arif JM, Lehmler HJ, Robertson LW, Gupta RC. Interaction of benzoquinone- and hydroquinone-derivatives of lower chlorinated biphenyls with DNA and nucleotides in vitro. *Chem. Biol. Interact.* 2003 ;142(3):307-16.

Borrelli I, Loffredo S, Staiano RI, Frattini A, Marone G, Triggiani M. Benzene metabolites inhibit the release of proinflammatory mediators and cytokines from human basophils. *Int. J. Immunopathol. Pharmacol.* 2010;23(3):737-44.

Callender VD, St Surin-Lord S, Davis EC, Maclin M. Postinflammatory hyperpigmentation: etiologic and therapeutic considerations. *Am. J. Clin. Dermatol.* 2011;12(2):87-99.

Chan K, Lehmler HJ, Sivagnanam M, Feng CY, Robertson L, O'Brien PJ. Cytotoxic effects of polychlorinated biphenyl hydroquinone metabolites in rat hepatocytes. *J. Appl. Toxicol.* 2010 ;30(2):163-71.

Chen C, Lai Y, Wu M, Zhang Z. (*Effects of DNA polymerase beta on certain genotoxicity of mouse fibroblast cell induced by hydroquinone).* Wei Sheng Yan Jiu. 2010 ;39(6):669-73.

Costa A, Moisés TA, Cordero T, Alves CR, Marmirori J. Association of emblica, licorice and belides as an alternative to hydroquinone in the clinical treatment of melasma. *An Bras. Dermatol.* 201085(5):613-20.

Galván N, Lim S, Zmugg S, Smith MT, Zhang LDepletion of WRN enhances DNA damage in HeLa cells exposed to the benzene metabolite, hydroquinone. *Mutat. Res*. 2008;649(1-2):54-61.

Grimes PE. Melasma. Etiologic and therapeutic considerations. *Arch. Dermatol.* 1995 ;131(12):1453-7.

Hariharan V, Toole T, Klarquist J, Mosenson J, Longley BJ, Le Poole IC. Topical application of bleaching phenols; in-vivo studies and mechanism of action relevant to melanoma treatment. *Melanoma Res*. 2011; 21 (2): 115-26.

Hu DL, Tang HW, Liang HR, Tang DS, Liu YM, Ji WD, Yuan JH, He Y, Zhu ZY, Yang JP, Fang DK, Sha Y, Tu XZ, Zhuang ZX. Possible role of DNA polymerase beta in protecting human bronchial epithelial cells against cytotoxicity of hydroquinone. *Biomed. Environ. Sci*. 2007 ;20(2):171-7.

Inayat-Hussain SH, Winski SL, Ross D. Differential involvement of caspases in hydroquinone-induced apoptosis in human leukemic hl-60 and jurkat cells. *Toxicol. Appl. Pharmacol.* 2001;175(2):95-103.

Ju L, Zhang SZ, Zhao R, Yao GD. *Differential proteomic expression in human liver cells stimulated by hydroquinone..* Zhonghua Lao Dong Wei Sheng Zhi Ye Bing Za Zhi. 200;24(11):658-61.

Kerzic PJ, Pyatt DW, Zheng JH, Gross SA, Le A, Irons RD. Inhibition of NF-kappaB by hydroquinone sensitizes human bone marrow progenitor cells to TNF-alpha-induced apoptosis. *Toxicology*. 2003 ;187(2-3):127-37.

Kim Y J, Woo HD, Kim BM, Lee YJ, Kang SJ, Cho YH, Chung HW. Risk assessment of hydroquinone: differential responses of cell growth and lethality correlated to hydroquinone concentration. *J. Toxicol. Environ. Health A.* 2009;72(21-22):1272-8.

Li X, Tang S, Huang H, Yang L, Liu J, Zhuang Z. Induction of a cell-survival adaptive response in MRC-5 cells by hydroquinone *Mutat. Res.* 200.8; 652(2):180-5.

O'Donoghue JL. Hydroquinone and its analogues in dermatology - a risk-benefit viewpoint. *J. Cosmet. Dermatol.* 2006 ;5(3):196-203.

Pandey AK, Gurbani D, Bajpayee M, Parmar D, Ajmani S, Dhawan A. In silico studies with human DNA topoisomerase-II alpha to unravel the mechanism of in vitro genotoxicity of benzene and its metabolites. *Mutat. Res.* .2009 ;661(1-2):57-70.

Pons M, Marin-Castaño ME. Cigarette smoke-related hydroquinone dysregulates MCP-1, VEGF and PEDF expression in retinal pigment epithelium in vitro and in vivo.

Ribas J, Schettini AP, Cavalcante Mde S. Exogenous ochronosis hydroquinone induced: a report of four cases. *An Bras. Dermatol.* 2010 ;85(5):699-703.

Rubio V, Zhang J, Valverde M, Rojas E, Shi ZZ. Essential role of Nrf2 in protection against hydroquinone- and benzoquinone-induced cytotoxicity. *Toxicol. In Vitro*. 2011;25(2):521-9.

Shi YF, Yu K, Chen Y, Ren XZ, Bi LX, Qian HL. *Effect of hydroquinone on expression of topoisomerase enzyme IIα in human bone marrow mononuclear cells*. Zhonghua Lao Dong Wei Sheng Zhi Ye Bing Za Zhi. 2010 ;28(9):660-3.

Shen DX, Shi X, Fu JL, Zhang YM, Zhou ZC. The role of thiol reduction in hydroquinone-induced apoptosis in HEK293 cells. *Chem. Biol. Interact.* 2003 ;145(2):225-33.

Topping DC, Bernard LG, O'Donoghue JL, English JC. Hydroquinone: acute and subchronic toxicity studies with emphasis on neurobehavioral and nephrotoxic effects. *Food Chem. Toxicol.* 2007 ;45(1):70-8.

Wan J, Winn LM. Benzene's metabolites alter c-MYB activity via reactive oxygen species in HD3 cells. *Toxicol. Appl. Pharmacol.* 2007;222(2): 180-9.

Westerhof W, Kooyers TJ. Hydroquinone and its analogues in dermatology - a potential health risk. *J. Cosmet. Dermatol.* 2005 1(2):55-9.

Wu XR, Xue M, Li XF, Wang Y, Wang J, Han QL, Yi ZC. Phenolic metabolites of benzene inhibited the erythroid differentiation of K562 cells. *Toxicol. Lett.* 2011;203:190-9.

Yang EJ, Lee JS, Yun CY, Kim IS. The pro-apoptotic effect of hydroquinone in human neutrophils and eosinophils. *Toxicol. In Vitro.* 2011 ;25(1): 131-7.

Zhang ZB, Hu J, Bi YY, Zhao ZW, Tao N, Yan H, Zhu ZL, Liu YM, Li L. *Effect of TTRAP expression on apoptosis induced by hydroquinone in HL-60 cells in vitro.* Zhonghua Lao Dong Wei Sheng Zhi Ye Bing Za Zhi. 2007 ;25(11):654-6.

Zhao Z, He X, Bi Y, Xia Y, Tao N, Li L, Ma Q. Induction of CYP4F3 by benzene metabolites in human white blood cells in vivo in human promyelocytic leukemic cell lines and ex vivo in human blood neutrophils. *Drug Metab. Dispos.* 2009;37(2):282-91.

INDEX

A

access, 66
accounting, 80, 86
acetaminophen, 139
acetic acid, 75, 76, 78, 85, 93, 94, 95, 96, 99
acetone, 64, 65, 66
acetylation, 173
acid, 11, 17, 18, 33, 40, 62, 63, 86, 89, 93, 107, 147, 148, 170, 178
acidic, 64, 65, 106, 139
acidity, 65
acquaintance, 46
activation state, 26
active site, 134
acute myeloid leukemia, 168, 181
adaptation, 176
adaptive immunity, 4
adenocarcinoma, x, 167
adhesion, 45, 46, 47
adhesives, 169
adsorption, 150
adults, 45
adverse effects, x, 167
age, vii, 1, 4, 6, 30, 36, 37, 38, 39, 40, 41, 42, 43, 44, 45, 48, 51, 52, 53, 54, 55, 170
age-related macular degeneration, vii, 1, 4, 36, 37, 38, 39, 40, 41, 42, 43, 45, 48, 51, 52, 53, 54, 55
aggregation, 4, 50
alcohols, 62, 63, 74, 75, 76, 83, 86, 94, 95
allergy, 4, 174
alters, 10
amines, 137
amino, 147
amino acid, 147
ammonium, 66, 121
amplitude, 127
anatomy, 5, 20
angiogenesis, vii, 2, 10, 12, 27, 28, 29, 38, 45, 51, 53, 54, 55, 179
aniline, 61, 63, 65
antibody, 16, 23, 42
antigen, 35, 49, 50
antimalarials, 169
antioxidant, 10, 34, 61, 147, 168, 179
antiozonants, 61
aplastic anemia, 168, 171, 181
apoptosis, ix, x, 4, 6, 12, 13, 35, 45, 46, 49, 55, 146, 147, 167, 171, 172, 173, 174, 175, 177, 179, 183, 184
aqueous solutions, 113, 150
Argentina, 103
argon, 157
ARM, 36
ascorbic acid, 139
assessment, 182, 183
asthma, 34, 45
atherosclerosis, 20
atmospheric pressure, 76
atoms, 137, 158

atrophy, 5, 6, 7, 13
Au substrate, 157
autoantibodies, 42
autolysis, 8
avian, x, 167

B

bacterial infection, 4
ban, 168, 181
basal lamina, 5, 8, 38, 45
base, 157
basement membrane, 5, 8, 9, 21, 24, 31
basic research, 107
basophils, 174, 182
BD, 40, 54
beer, 2
benzene, ix, x, 2, 3, 4, 33, 34, 35, 63, 64, 86, 98, 145, 167, 168, 169, 171, 172, 173, 174, 175, 178, 181, 182, 183, 184
Benzene, ix, 145, 171, 172, 174, 179, 182, 184
benzoquinone (BQ), ix, 146, 150, 153, 154, 155, 158, 161, 174, 178, 179
binding energy, 157, 160
bioaccumulation, 63
bioassay, 181
biochemical processes, 147
biological activity, 48
biomarkers, 37, 42, 150, 161
biomolecules, 7
biosensors, 106, 107, 139
biosynthesis, 147
bisphenol, 66
bleaching, 169, 170, 182
blindness, vii, 1, 4, 32
blood, 4, 6, 7, 10, 22, 26, 27, 30, 33, 147, 171, 173, 184
blood flow, 10
blood supply, 6
blood vessels, 7, 22, 26, 28
bonding, 59, 62, 70, 73, 76, 81, 86, 91
bonds, 14, 114
bone, ix, 4, 34, 35, 145, 168, 173, 181, 183
bone marrow, ix, 4, 34, 145, 168, 173, 181, 183
brain, 50
breakdown, 9, 21, 52
bronchial asthma, 174
bronchial epithelial cells, 180, 182
by-products, 65, 66

C

Ca^{2+}, 47, 174
cadmium, 10, 12, 40
cancer, 4, 22, 51, 168, 181
capillary, 33
carbon, viii, 58, 62, 68, 69, 70, 71, 72, 73, 74, 91, 92, 99, 146, 150, 160
carbon dioxide, viii, 58, 62, 68, 69, 70, 71, 72, 73, 74, 91, 92, 99
carboxylic acid, 62
carboxylic acids, 62
carcinogen, 179, 181
carcinogenesis, 4, 168, 181
carcinogenicity, 33, 181
carotene, 41
carotenoids, 10
cascades, 18
caspases, 174, 183
catalase, ix, 146
catalysis, 65
catalyst, 65
catalytic effect, 107
catalytic hydrogenation, 65
cation, 106, 114, 115, 116, 121
cell death, 6, 12, 13, 45, 48, 49, 172, 177, 179
cell line, 17, 50, 171, 172, 180, 184
cell lines, 184
cell signaling, 176
cellulose, 33
central nervous system, 52
charge density, 120
chemical, viii, ix, 3, 11, 33, 57, 59, 61, 62, 67, 80, 82, 83, 89, 107, 146, 150, 154, 161, 170, 173
chemical engineering, viii, 57, 67

chemical industry, 62, 67
chemical peel, 170
chemical properties, 67, 80
chemical reactions, 154
chemical reactivity, 3
chemicals, ix, 2, 61, 67, 145, 174
chemisorption, 113
chemokines, 26, 28
chemotaxis, 46
chicken, 50, 171
children, 147
China, 167
cholesterol, 11
choroid, vii, 2, 11, 12, 16, 26, 31, 43, 55
choroidal neovascularization (CNV), vii, 2, 7, 178
chromatography, 33
chromatography analysis, 33
chromosome, 10, 41, 178
chronic renal failure, 147
cigarette smoke, vii, ix, 1, 2, 4, 8, 10, 11, 12, 13, 28, 29, 32, 33, 34, 35, 40, 41, 43, 44, 55, 145, 179
cigarette smokers, 4, 41
cigarette smoking, 10, 13, 39, 40, 43
classification, 36
cleaning, 152, 157
cleavage, 14, 20, 174, 175
clinical presentation, 5
CNS, 180
CO_2, 70, 72
coffee, 2, 63
collagen, 20, 21, 49
color, vii, 148, 168
commercial, 66, 153
community, 32, 148
complement, 6, 9, 11, 12, 25, 41, 43
complex interactions, 49
complexity, 27, 81
complications, 10
composition, viii, 37, 57, 60, 67, 82, 92, 158
compounds, ix, 2, 3, 11, 61, 62, 63, 67, 74, 83, 92, 93, 97, 99, 105, 146, 149, 150, 151, 160, 161, 163, 168, 169, 181
computational modeling, 67
conductance, 137
conduction, 104, 106, 107, 123, 137, 139
conductivity, viii, 104, 105, 106, 107, 127, 134, 137, 139
configuration, 66, 133, 135, 136
conjugation, 34
consensus, 6, 7
conservation, 49
constituents, 8, 44, 69, 147
construction, 64
consumption, 41
contaminant, ix, 145
contamination, 160
control group, 180
COOH, 86
cooling, 61, 67, 76, 99
cornea, 4
correlation, 20, 39, 92, 93, 97, 99, 178
correlations, 39, 80, 88, 89, 97
corrosion, 105
cortex, 4
cosmetics, 61, 168, 169, 170, 181, 182
cost, 67, 95
crystal structure, 68
crystalline, 2, 76
crystallization, 61, 67, 74, 76, 99
crystals, 76
CT, 163, 178
Cubic-plus-Association equation of state (CPA), viii, 58
culture, 26, 28
CV, 106, 107, 116, 138, 152, 153, 157
Cyclic Voltammetry, viii, 104, 106, 107, 153
cycling, 3, 152, 157, 177
cystathionine, 148
cysteine, 3, 10, 41, 146, 147
cystine, 41
cytochrome, ix, 2, 4, 145, 177
cytochrome p450, 2
cytokines, 7, 9, 25, 26, 27, 38, 169, 174, 182
cytokinesis, 178
cytometry, 178
cytoplasm, 3, 8, 14, 15, 147
cytoskeleton, 2, 14, 15, 18, 20, 45, 46, 47

cytotoxicity, 169, 176, 179, 182, 183

D

damages, 3
data set, 81
database, 83, 84, 88, 89, 93
decomposition, 62
defence, 177
deficiency, 51, 53, 147, 179
degradation, ix, 8, 20, 21, 49, 99, 104, 107, 138, 139
dendritic cell, 10, 44, 52
dendritic spines, 46
dephosphorylation, 18
deposition, 4, 9
deposits, vii, 1, 2, 5, 6, 9, 11, 13, 21, 24, 25, 26, 27, 30, 31, 32, 38, 52
depth, vii, 2, 10
derivatives, 69, 150, 170, 174, 180, 182
dermatology, 183, 184
dermatosis, 169
dermis, 168, 169
destruction, 33, 35
detachment, 36, 39
detectable, 25, 30
detection, vii, 147, 148, 149, 160
detoxification, 39, 147
deviation, 59, 92, 93, 95
diabetes, 147
diet, 30, 31, 32, 35, 38
dietary fat, 49
diffusion, ix, 104, 105, 107, 108, 110, 112, 113, 114, 119, 120, 124, 127, 128, 129, 134, 135, 136, 137, 138, 139
diffusion process, 112, 137, 139
diffusion rates, ix, 104, 139
dimerization, 17
dioxin, 10, 13, 40
direct action, 12
disease progression, 6, 26
diseases, 4, 15, 26, 147, 148, 173, 174
disorder, 175
dissociation, 149, 177
distillation, 61, 67, 69, 91, 98
distillation processes, 67
distribution, viii, 57, 81, 137, 174
diversity, 168, 181
DNA, ix, x, 3, 12, 33, 34, 43, 44, 146, 167, 168, 175, 176, 178, 180, 181, 182, 183
DNA damage, 33, 34, 168, 176, 178, 180, 181, 182
DNA polymerase, 176, 180, 182
DNA repair, 43, 178
doctors, 168, 181
donors, x, 15, 16, 146, 149, 161, 173, 174
dosing, 148
down-regulation, 43
drawing, 5, 15
drinking water, 30, 31
drugs, 2, 35, 148, 170
drusen, vii, 1, 5, 6, 7, 8, 9, 11, 12, 13, 21, 25, 27, 30, 32, 37, 39, 45, 178
drying, 61, 67, 76
dyes, 169

E

ECM, vii, 2, 9, 13, 20, 21, 22, 24
edema, 147
EIS, 106, 107, 124, 138
Elam, 163
elderly population, 178
electrocatalysis, 105, 106
electrochemical behavior, 150
Electrochemical Impedance Spectroscopy, viii, 104, 106, 107
electrochemistry, 105, 107, 139, 153, 157
electrode surface, 113
electrodes, vii, viii, 103, 104, 105, 106, 107, 124, 139, 150, 153, 157, 158
electrolyte, 105, 112, 113, 117, 120, 124, 128, 131
electron, viii, ix, 3, 8, 31, 82, 104, 105, 107, 108, 114, 117, 119, 123, 124, 127, 128, 132, 133, 134, 135, 136, 138, 139, 146, 149, 150, 153, 154, 156
electron microscopy, 8, 31
electrons, 112, 113, 114, 117, 127, 129, 153
electrophoresis, 34, 176

EMMPRIN, 20, 50, 51
encoding, 171, 172
endothelial cells, 12, 22, 28, 47, 49
endothelium, 30, 31
energy, 58, 60, 63, 67, 68, 84, 85, 86, 87, 91, 105, 135, 157, 176, 177
energy consumption, 67
engineering, viii, 57, 67
England, 140
entropy, 60, 78, 79, 82
environment, 30, 63
environmental stimuli, 18
environments, 7
enzyme(s), ix, 8, 20, 39, 40, 43, 47, 146, 148, 172, 178, 183
eosinophils, 173, 184
epidermis, 168
epithelial cells, 37, 40, 42, 43, 45, 48, 52, 53, 54, 55, 179, 180
epithelium, vii, 1, 5, 16, 37, 38, 39, 42, 43, 45, 48, 49, 53, 54, 55, 178, 183
epitopes, 42
equilibrium, viii, 53, 57, 59, 63, 67, 68, 69, 74, 80, 81, 86, 92, 96, 110
erythrocytes, 147
ester, 62, 86, 95
estrogen, 49
ethanol, 76, 78, 79, 94, 95, 153, 157
ethers, 74
ethnic groups, 168
ethyl acetate, 76, 78, 79, 94, 96
etiology, 10, 40, 42, 170
evidence, vii, 1, 3, 8, 15, 20, 22, 25, 26, 28, 29, 106, 113, 168, 175, 181
exchange rate, 135
exclusion, 120
exocytosis, 46
exposure, 2, 4, 6, 8, 13, 17, 27, 28, 29, 30, 33, 34, 35, 38, 45, 49, 170, 171, 172, 175, 176, 177, 179, 181
extracellular matrix, 2, 6, 9, 20, 37, 48, 49, 50, 51
Extracellular matrix turnover, vii, 2
extraction, 67, 92, 99
extracts, 41, 63
extrusion, 9

F

families, 47
fasting, 148
fat, 30, 31, 32
fear, 32
feed composition, viii, 57, 67
fiber, 47
fibroblasts, x, 50, 52, 167, 176
filament, 14, 46, 47
film thickness, 106, 107, 109, 110, 111, 112, 113, 117, 127, 129, 139
films, viii, 104, 106, 108, 110, 113, 114, 115, 116, 117, 118, 119, 120, 121, 124, 127, 129, 130, 131, 134, 135, 136, 137, 138, 139
flexibility, 65
flora, 36
Flory-Huggins approximation, 82
fluid, viii, 57, 59, 60, 62, 70, 71, 81, 88, 91, 99
fluid extract, 99
fluorescence, 17, 171
food, 2, 30, 61, 62, 63, 67, 98
Food and Drug Administration, 170
food products, 63
formation, 5, 6, 7, 8, 9, 10, 12, 13, 17, 18, 19, 20, 21, 24, 25, 27, 28, 31, 32, 33, 36, 37, 38, 45, 47, 49, 66, 91, 147, 150, 157, 173, 176
formula, 149
fragments, 8, 137
France, 57
free radicals, 10, 11, 26, 33, 41, 61
fruits, 2, 63
fusion, 59, 63, 68, 79

G

GDP, 177
gel, 34
gelatin zymography, 24

gelatinase A, 50
gene expression, 21, 27, 38, 52, 171
gene regulation, 47
genes, 11, 21, 33, 39, 41, 172, 173, 179
genome, 36
genomic stability, 175
genotype, 11, 27
Gibbs energy, viii, 57, 59, 62, 78, 81, 93, 95
glucose, 106
glutathione, ix, 10, 12, 43, 146, 148, 168, 177, 179, 181
glycol, x, 167
glycosylation, 21, 50
GPA, 172
grading, 36
grants, 32
granules, 8
graph, 24
group interactions, 83
growth, vii, 1, 7, 9, 10, 12, 27, 44, 45, 53, 54, 55, 112, 183
growth factor, 7, 9, 10, 12, 28, 44, 45, 53, 54, 55
GTPases, 46
guanine, x, 167

H

hair, 169
harmful effects, 178
health, iv, vii, 6, 9, 21, 25, 37, 168, 181, 184
health effects, iv, vii
health risks, 168, 181
heat capacity, 68
heat shock protein, 14, 45, 46, 47, 48
hematopoietic system, 172
heme, 172
hemoglobin, 172
hepatocytes, x, 167, 177, 182
heterogeneity, 26
heterogeneous catalysis, 65
high fat, 38
histamine, 174
histone, 173
history, 16, 43
homeostasis, 29, 30, 147
homocysteine, 146, 147
Hong Kong, 167
hormones, 9, 25
HPC, 172
human, x, 2, 4, 15, 16, 17, 20, 22, 23, 25, 26, 27, 29, 30, 34, 35, 37, 39, 40, 41, 42, 43, 44, 45, 48, 49, 50, 52, 53, 54, 55, 146, 147, 167, 169, 171, 173, 174, 176, 177, 178, 179, 180, 181, 182, 183, 184
human exposure, 2
human health, 4
human neutrophils, 171, 173, 184
hydrogen, 59, 61, 62, 63, 64, 65, 70, 73, 76, 81, 82, 86, 91, 114, 137, 148, 152, 157, 177
hydrogen bonds, 82, 114
hydrogen peroxide, 61, 64, 65, 148, 177
hydrolysis, 66
hydrophilicity, 83
hydrophobicity, 83
Hydroquinone, i, iii, v, vi, vii, ix, x, 1, 2, 3, 4, 13, 25, 30, 34, 35, 57, 61, 63, 68, 73, 86, 90, 100, 103, 107, 145, 167, 168, 169, 170, 172, 173, 174, 181, 183, 184
hydroxyl, 3, 33, 61, 62, 63, 70, 76, 85, 86, 90, 98
hydroxyl groups, 3, 61, 63, 70, 86, 90, 98
hypothesis, 6, 7, 8, 9, 13, 25, 26, 27, 30, 32, 37, 172
hypoxia, 10
hysteresis, ix, 104, 134, 136

I

ICAM, 50
ID, 34
ideal, 58, 76, 77, 78, 80, 81, 91
identity, 14
IFN, 4, 35
IL-8, 173
immobilization, 150
immobilized enzymes, 139
immune function, 169
immune response, 3, 25, 34, 169, 173

immune system, 10, 25
immunocompetent cells, 52
immunofluorescence, 22
immunoglobulin, 35, 49, 50, 51
immunoglobulin superfamily, 49, 50
immunostimulatory, 26
impurities, 76
in utero, x, 167
in vitro, 21, 29, 30, 34, 38, 40, 44, 53, 148, 169, 171, 175, 178, 179, 182, 183, 184
in vivo, 12, 21, 29, 38, 48, 53, 169, 171, 179, 183, 184
incidence, 7, 10, 34, 39, 41, 99, 169
individuals, 172
inducer, 20, 22, 40, 50, 51
induction, 21, 169, 171, 176, 178
industrial chemicals, 61
industries, 62, 67, 80
industry, 2
infection, 4, 34
inflammation, vii, 2, 3, 6, 11, 12, 25, 34, 37, 43, 44, 147, 174
inflammatory disease, 52
inflammatory mediators, 10
inflammatory responses, 11, 26, 34
ingestion, 2, 53
inhibition, 18, 137, 172, 178
inhibitor, 14, 18, 19, 28, 47, 48, 55, 61, 168, 174, 175, 177
initiation, 11, 29
injections, 31
injury, vii, 2, 6, 7, 8, 9, 10, 12, 13, 14, 15, 17, 20, 21, 26, 27, 28, 32, 37, 38, 46, 49, 179
innate immunity, 4
insulators, 104
integration, 115
integrin, 53
integrity, ix, 146
interface, 25, 37, 107, 110, 112, 114, 117, 128, 131, 133, 134, 137, 138, 139
interference, 139, 175, 180
interferon, 4
intermolecular interactions, 67
ion transport, 127, 136, 137, 138
ions, 47, 105, 114, 119, 137
iron, 65, 66, 114, 119
isolation, 33
isomers, viii, 58, 61, 63, 68, 69, 70, 91, 92, 93
isotherms, 92

K

kidney, 2, 4, 147, 177
kill, 6, 9
kinase activity, 18
kinetics, 18, 150

L

lactate dehydrogenase, 176
laminar, 5, 49
Langerhans cells, 169
lasers, 170
LDL, 11
lead, viii, ix, 9, 20, 27, 28, 29, 30, 80, 104, 114, 120, 135, 146, 168, 180
leakage, 176
leaks, 3
legal blindness, 178
lens, 147
lesions, 25, 53, 168, 180
leucocyte, 147
leukemia, 4, 36, 169, 171, 174
leukocytes, 147, 171, 173
leukomogenicity, x, 167
LFA, 50
light, 7, 8, 27, 30, 32, 38, 49, 53, 99, 170
linear function, 92
linear model, 80
lipid metabolism, 50
lipid oxidation, 41
lipid peroxidation, 3, 12
lipids, 11, 12
lipoproteins, 38
liquid phase, 91
liquids, 68
liver, 2, 4, 33, 35, 147, 168, 177, 181, 183

liver cells, 183
localization, 42, 54
loci, 36
locus, 11
lung disease, 20, 40
lymphocytes, 4, 34, 35, 44

M

macrophages, 4, 5, 6, 9, 10, 25, 26, 27, 34, 36, 38, 49, 169, 179
macular degeneration, vii, 1, 4, 9, 36, 37, 38, 39, 40, 41, 42, 43, 44, 45, 48, 51, 52, 53, 54, 55, 178
magnitude, 30, 70, 73, 105, 110, 133, 135, 138
majority, 67, 80, 99, 147
malignancy, 50
malnutrition, 147
mammalian cells, 45, 46, 178
management, 32
manganese, 65, 66
manufacturing, 2, 69
marrow, 4, 168, 181
mass, 114, 176
mass spectrometry, 176
materials, 61, 67, 99, 106, 137
matrix, vii, 2, 9, 20, 21, 49, 50, 51
matrix metalloproteinase, 9, 20, 49, 50, 51
matter, iv, 106
MCP, 9, 26, 27, 32, 52, 53, 173, 178, 183
MCP-1, 9, 26, 27, 32, 52, 53, 173, 178, 183
measurement, x, 129, 136, 146, 161
measurements, 62, 67, 70, 76, 81, 97, 124, 129, 135, 138, 153, 157
media, 64, 99, 106, 139, 149
mediation, viii, 104, 106, 114, 139
medical, 61, 170
MEK, 18
melanin, 8, 168
melanoma, 169, 182
melting, 60, 62, 68, 73, 79, 80, 83
melting temperature, 68, 79
membranes, 3, 28, 36, 52, 54
menadione, 45
metabolic pathways, 148
metabolism, 3, 8, 11, 33, 35, 147, 176, 177
metabolites, ix, 33, 34, 35, 145, 171, 172, 173, 174, 175, 177, 178, 179, 180, 182, 183, 184
metabolized, ix, 2, 145, 171
metalloproteinase, 20, 50, 51
methanol, 74, 76, 78, 93, 94, 95, 96
methodology, 81, 99
methyl methacrylate, 35
Miami, 1
mice, 18, 27, 29, 30, 33, 35, 49, 51, 53, 55, 169, 179
Michael acceptors, x, 146, 149, 161
Michael donors, x, 146, 149, 161
Michael type. Quinones, x, 146, 161
micronucleus, 176, 178
microscope, 17, 23
microscopy, 16, 19
microsomes, 33, 35
migration, 4, 9, 21, 26, 28, 127, 169
mitochondria, 3, 147
mitochondrial DNA, 33, 37, 43
mitogen, 14, 47
mixing, 74, 82, 87
MMP(s), 9, 20, 21, 22, 23, 24, 32, 37, 49, 50, 51
MMP-2, 9, 20, 21, 22, 24, 32, 37, 50
MMP-9, 49
models, viii, 8, 28, 30, 57, 62, 63, 80, 81, 91, 93, 95, 104
modifications, 17, 80, 84, 150
molar volume, 91
mole, 58, 60, 70, 78, 79, 82, 87, 92
molecular mass, 19
molecular structure, 91, 99
molecular weight, 21, 58, 92
molecules, x, 9, 13, 14, 20, 21, 24, 27, 29, 61, 62, 76, 80, 81, 82, 83, 84, 86, 87, 89, 90, 91, 92, 99, 146, 149, 153, 154, 160, 161, 179
monocyte chemoattractant protein, 9, 26, 52, 178
monolayer, 150
monomers, 14

morphology, 30
mortality, 180
mRNA(s), x, 26, 49, 167, 171, 172, 173, 175, 179
mtDNA, 12
mutations, 168, 175, 181

N

NADH, 34
National Academy of Sciences, 42
necrosis, 175, 176
neovascularization, vii, 2, 5, 7, 9, 12, 38, 40, 44, 52, 54, 55, 178
nervous system, 4
neurotoxicity, 147
neutral, 106, 107, 139
neutrophils, 173, 184
nicotine, 10, 12, 44
nitric oxide, 3, 33, 34
nitrogen, 10, 114, 137, 153, 158
nitrogen dioxide, 10
nitrogen gas, 153, 158
Non-Random Two Liquid Segment Activity Coefficient model (NRTL-SAC), viii, 58
novel materials, 105
Nrf2, 179, 183
nucleation, 46
nuclei, 8, 16
nucleophiles, x, 146, 149, 161
nucleotides, 182
nucleus, 2, 8, 47, 179
null, 51, 176

O

observable behavior, 180
ocular diseases, 40
oligomers, 15
omission, 16, 23
operations, 67
optimization, 67
organ, 6, 13
organelles, 8, 13
organic compounds, 67, 70, 146
organic solvents, viii, 58, 78, 95
osteoporosis, 147
ox, 128
oxidation, 3, 10, 14, 15, 38, 41, 42, 61, 63, 65, 105, 106, 108, 110, 112, 113, 114, 120, 146, 148, 153, 154, 156, 177
oxidative damage, x, 10, 11, 12, 13, 15, 27, 28, 29, 30, 40, 55, 167, 176, 180
oxidative stress, ix, 6, 8, 10, 12, 14, 15, 18, 20, 27, 28, 29, 34, 38, 40, 43, 45, 46, 47, 146, 147, 177, 179
oxygen, 12, 114, 152, 157, 160, 173

P

pairing, 114, 139
parameter estimation, 62
partition, 80, 110, 113
pathogenesis, vii, 1, 4, 6, 8, 9, 10, 13, 14, 15, 20, 26, 28, 30, 36, 37, 38, 43, 174
pathology, 21, 43, 45, 48
pathophysiology, 11, 40
pathways, 6, 17, 18, 20, 32, 45
PCBs, 177
PCR, 49, 172
perchlorate, 113
peripheral blood, 171
permeability, 53, 113
permeation, viii, 104, 107, 110, 112, 113, 114, 120, 139
permission, iv
permit, 22, 29
peroxidation, 12
peroxide, 45, 65
peroxynitrite, 10, 34
pH, 106, 113, 129, 130, 139, 149, 153, 154, 155
phagocytosis, 27, 42, 179
pharmaceutical, 62, 67, 80
pharmaceuticals, 61
phase equilibria, viii, 57, 62, 67, 80, 81, 99
phase transitions, 79

phenol, ix, 33, 35, 61, 63, 64, 65, 66, 68, 69, 70, 86, 89, 91, 95, 96, 97, 99, 145, 169, 171, 172
phenolic compounds, 89, 98
phenotype, 37
phenylalanine, 174
phosphatidylserine, 175
phospholipids, 42
phosphorylation, 14, 15, 17, 18, 19, 32, 46, 47, 48, 174
photoemission, 158
photographs, 7
physical interaction, 62, 86
physiology, 20, 22
placebo, 41
plant growth, 66
plants, 63
plasma membrane, 13, 14, 17, 22
platinum, 152
poison, ix, 146
polar, 58, 82, 83
polarity, 83
polarized double bond, x, 146, 149, 161
pollutants, 177
pollution, 8
polychlorinated biphenyl, 180, 182
polychlorinated biphenyls (PCBs), 180
polymer, viii, 81, 104, 105, 106, 107, 108, 110, 112, 113, 114, 115, 116, 117, 119, 124, 127, 128, 129, 130, 131, 132, 133, 134, 136, 137, 138, 139
polymer chain, 137
polymer chains, 137
polymer electrolytes, 105
polymer film, viii, 104, 105, 107, 108, 110, 112, 113, 115, 127, 130, 136, 137, 139
polymer films, 105, 136, 139
polymer matrix, 136, 139
polymer structure, 112
polymerase, 180
polymeric materials, 105, 138
polymerization, 14, 17, 47, 61, 168
polymers, 61, 104, 105, 136, 137
polymorphism(s), 11, 43, 175
population, 5, 32, 36
Portugal, 57
pregnancy, 170
preparation, 98, 105, 150
preservation, 21
prevention, 32, 48
primate, 45
progenitor cells, 172, 183
pro-inflammatory, 9, 11, 12, 44, 178
proliferation, 28, 34, 44, 176
propagation, 136, 140
prophylactic, 169
protection, 10, 40, 105, 179, 183
protective mechanisms, 168, 181
protective role, 44
protein folding, 15, 176
protein oxidation, 3
protein synthesis, 147
proteins, ix, 3, 10, 11, 12, 14, 15, 21, 24, 32, 46, 47, 48, 146, 147, 176, 179
proteoglycans, 37
proteolysis, 51
proteome, 42
protons, 105, 119, 120, 137
psoriasis, 147
public health, 5, 6
Puerto Rico, 145
purification, viii, 57, 67, 74, 98, 99

Q

quality of life, 168
quantification, 148, 178
quinone, 2, 12, 71, 73, 74, 92, 148, 174, 177
quinones, ix, x, 3, 10, 12, 35, 146, 149, 161, 167

R

radial distribution, 58, 88
radicals, 3, 10, 26, 33, 41
reactants, 110
reactions, 52, 99, 105, 146, 148, 150
reactive oxygen, ix, 2, 3, 35, 146, 171, 174, 179, 184

reactive oxygen species (ROS), ix, 3, 146, 171, 174
reactivity, 3, 34, 149
receptors, 12, 54
recombination, 34
reconstruction, 52
recovery, 68, 70, 98
reform, 14
regression, 36, 63, 78, 81, 89, 93, 129
relaxation, 136
relevance, 21, 35, 43
reliability, 67
repair, 14, 33, 176
reproduction, 36
repulsion, 136
requirements, 67
residuals, 129
residues, 3
resistance, 107, 127, 128, 132, 133, 138
resolution, 36, 157, 160
resorcinol, 3, 61, 63, 64, 65, 70, 71, 74, 76, 77, 78, 86, 89, 92, 97, 98, 99, 100, 169
response, 2, 8, 11, 13, 14, 15, 18, 20, 25, 27, 29, 32, 46, 114, 115, 121, 127, 150, 169, 172, 175, 176, 178, 179, 183
retina, vii, 1, 4, 5, 6, 7, 12, 13, 16, 22, 23, 26, 29, 31, 37, 40, 41, 45
retinal disease, 12
ribonucleotide reductase, 34
risk, vii, 1, 4, 6, 7, 10, 11, 26, 32, 38, 41, 43, 52, 147, 148, 170, 175, 178, 183, 184
risk factors, 6
risks, 147
RNA, 54, 175, 179, 180
RNAi, 175
rodents, 181
Rotating Disc Electrode Voltammetry, viii, 104, 106, 107
routes, 61, 63, 98
rubber, 2
rules, 87

S

safety, 35, 170, 182
scaling, 129
scattering, 76, 78
secrete, 26
secretion, 7, 8, 9, 10, 27, 28, 53, 54
seedlings, 63
self-assembly, 150
semicircle, 125, 127
sensors, 105, 106, 139
serum, 11, 19, 43
sex, 181
shock, 14, 15, 46, 47, 48
showing, 16, 17, 24, 25, 27, 63, 107
signal transduction, 47, 177
signaling pathway, ix, 18, 47, 146, 171, 172
signalling, 54
simulations, 129
siRNA, 18, 28, 54
skeleton, 177
skin, vii, x, 2, 4, 35, 61, 167, 168, 169, 170, 181
smoke-related hydroquinone, vii, 1, 4, 28, 32, 48, 53, 183
smoking, vii, 1, 10, 11, 32, 37, 40, 41, 42, 43
smoking cessation, vii, 1
smooth muscle, 12, 44, 45
smooth muscle cells, 12, 44
society, 5
sodium, 66, 153
sodium hydroxide, 66, 153
solid phase, viii, 57, 79
solid surfaces, 150
solubility, vii, 58, 61, 62, 67, 70, 71, 72, 73, 74, 76, 77, 78, 79, 80, 83, 91, 92, 93, 94, 97, 99
solvation, 92, 95
solvents, 61, 62, 63, 65, 67, 68, 75, 76, 78, 79, 83, 86, 89, 93, 94, 95, 97, 99
species, ix, x, 3, 12, 29, 35, 50, 59, 60, 82, 85, 104, 106, 107, 110, 112, 113, 114, 117, 120, 121, 124, 127, 128, 129, 139, 146, 148, 149, 153, 167, 171, 173, 174, 179, 181, 184
spectrophotometric method, 148
spectroscopy, 138

spleen, 4, 147
stability, 47, 140, 150, 161
standard deviation, 129
standard error, 24
state, viii, 46, 58, 59, 62, 79, 80, 81, 92, 93, 99, 105, 108, 109, 114, 118, 128, 135, 136, 137
states, 114
statistics, 78
stimulation, 4, 26, 47, 50, 180
stimulus, 8, 30, 32
storage, 61, 67, 105
stress, ix, 11, 14, 29, 37, 42, 46, 47, 48, 146
structure, ix, 3, 14, 47, 62, 63, 76, 82, 83, 89, 129, 146, 150, 153, 161, 176
substitution, 89, 91
substrate(s), 3, 129, 130, 131, 152, 153, 157, 158
sulfate, 66, 169
sulfur, x, 146, 149, 161
sulfuric acid, 65
Sun, 40, 42, 50, 162
superoxide dismutase, ix, x, 146, 167
supplementation, 41
suppression, 4
surface area, 82
surfactants, 150
survival, 28, 172, 176, 183
susceptibility, 11, 33, 36, 41, 43, 175, 177, 179
swelling, 120, 136
symptoms, 148
syndrome, 175
synthesis, ix, 9, 12, 24, 44, 61, 105, 145, 172, 174, 177

T

T cell, x, 4, 35, 167, 169, 175
tar, 10, 34, 35, 41, 63
target, 3, 6, 32, 45, 169, 176
TBP, 169
techniques, viii, 99, 104, 105, 106, 107, 148
technology, 70
TEM, 30, 31
temperature, viii, 57, 58, 67, 68, 70, 71, 72, 73, 75, 76, 77, 78, 79, 81, 87, 88, 89, 91, 92, 93, 94, 95, 96, 97, 98, 99
testing, 30
TGF, 53
theoretical approaches, 62
therapeutic interventions, 30
therapy, 170
thermodynamic properties, 76, 78
thermodynamics, viii, 57, 67
thin films, 113
TIMP, 20, 49
tissue, 6, 10, 11, 13, 20, 22, 25
titanium, 65
TNF, 26, 172, 183
TNF-alpha, 172, 183
TNF-α, 26
tobacco, 32, 33, 45, 63
tobacco smoke, 45, 63
tobacco smoking, 32
toxic effect, 173, 175
toxic substances, 7
toxicity, ix, 2, 3, 35, 43, 145, 148, 161, 168, 169, 172, 173, 174, 175, 177, 180, 181, 184
toxicology, 35
transcription, 14, 15, 51, 171, 172, 173, 179
transcription factors, 14, 15
transference, 127
transforming growth factor, 53
transmembrane glycoprotein, 21
transport, vii, viii, 3, 34, 46, 103, 104, 105, 106, 108, 114, 116, 117, 119, 124, 130, 134, 135, 137, 138
transport processes, 105, 106, 114, 137
transportation, 61, 67, 76
treatment, 10, 18, 20, 115, 116, 121, 123, 124, 134, 139, 148, 152, 157, 169, 170, 172, 173, 176, 179, 182
trial, 41
tumor, 50, 51, 169
tumor cells, 51, 169
tumor growth, 169
tumors, 169

turnover, vii, 2, 9, 13, 14, 15, 20, 21, 24, 37, 49

U

ulcer, 35
ultrastructure, 39
United States US, 5, 42, 61, 100, 142
urban, 7
uric acid, 106, 139
urine, 2, 33, 147
UV, 170
UV radiation, 170
UVB irradiation, 48

V

vapor, 59, 63, 65, 67, 68, 69, 70, 73, 86, 88, 89, 90, 91, 96, 98, 99
variables, 80, 106
variations, 11, 116
vascular endothelial growth factor (VEGF), 10, 12, 40, 55, 178
vascularization, 10
vector, 180
vegetables, 63
VEGF expression, 28, 29
VEGF protein, 179
VEGFR, 55
Venezuela, 145
vesicle, 14
vessels, vii, 1, 6
victims, 171
viscosity, 114, 128
vision, vii, 1, 4, 5, 41
vitamin C, 10, 41
vitamin E, 49
vitamins, 41
vitiligo, 168, 169, 181

W

waste, 66, 70
water, vii, viii, 2, 58, 62, 63, 66, 74, 75, 76, 78, 79, 80, 83, 86, 93, 94, 95, 97, 98, 99, 154, 157
Western blot, 16, 19, 22
wheat germ, 63
white blood cells, x, 167, 171, 184
wild type, 169
workers, 35, 181
World Health Organization, 100
worldwide, 5
wound healing, 53

X

XPS, 146, 152, 157, 158, 159, 160
X-ray photoelectron spectroscopy (XPS), 150, 157, 158

Y

yield, 66

Z

zinc, 41